12 Volt Electrics & Solar Power
Made Simple

Nathan Savage

Copyright©2017 Nathan Savage

ISBN-13: 978-1724783646

ISBN-10: 1724783645

Produced in the United Kingdom

All rights reserved. No part of this publication may be reproduced, stored in a retrieval system, or transmitted in any form or by any means, electronic, mechanical, photocopying, recording or otherwise without written permission of the author or publisher.

Disclaimer

Although the author and publisher have made every effort to ensure that the information in this book was correct at the time of publication, the author and publisher do not assume and hereby disclaim any liability to any party for any loss, damage, or disruption caused by errors or omissions, whether such errors or omissions result from negligence, accident, or any other cause.

Contents

Chapter 1 About this book	1
Chapter 2 The basic components needed	2
Chapter 3 Buying the right battery	3
Chapter 4 Choosing a solar panel	9
Chapter 5 Creating a basic safe circuit	14
Chapter 6 Connecting things together	16
Chapter 7 Fuse boxes & keeping it safe	21
Chapter 8 The switch	25
Chapter 9 Light bulbs	29
Chapter 10 Selecting the right wire for your circuit	32
Chapter 11 Inverters	35
Chapter 12 Installing it & connecting it all up	37
Chapter 13 Series & parallel circuits	40
Chapter 14 Goodbye	47

Chapter 1

About this book

This book is designed to help the absolute beginner who has no experience with working within the field of 12 volt and solar power. The information in this book is designed for those of you who simply wish to fit a very basic off grid power system into a shed, small boat or a van. This book contains all the knowledge & information that you need to get the job done very quickly and effectively.

There is no steep learning curve involved in fitting this because a very simplified system is used to enable you to create a very safe and fast installation. Most solar & 12 volt books will require you to learn about various types of circuits and how they are constructed and calculated for their amp rating and voltage drop problems, which is information that a lot of people struggle with. This book does away with all of this because all the calculations are done for you. This means that all you have to do is cut your wiring to the right length and connect up your 12 volt devices, it is actually that simple.

Starting with the basics of batteries we shall move onto how to connect your own solar panel, create a basic super safe circuit and how to fit power sockets and lights into your shed, boat, van etc. You will then be showed how you can easily power ordinary mains powered devices such as televisions, vacuum cleaners, power tools, game consoles, mobile phones and so much more quickly and easily from your 12 volt power supply.

This new simplified system means that you can have any small shed, boat or van completely wired up and off grid in less than a day even if you have no knowledge of electrical circuits whatsoever.

If you are a beginner in the field of 12 volt & solar panels then this book is written for you. It won't show you how to create a huge monolithic solar array that will supply your entire household with power but it will show you everything that you need in order to provide your own off grid power needs for man caves, sheds, vans and small boats. It will also give you all the basic knowledge that you need as a beginner for you to learn how to build & improve on this design in order to build bigger and more powerful systems in the future.

Chapter 2

The Basic Components needed

To build this super simple installation you will require some basic components that are readily available to buy from eBay, Amazon and a host of other online retailers. The items on this list will be covered throughout this book showing you how they are used to create a basic 12 volt off grid system for all of your basic power needs.

A 100 watt solar panel

A 30 amp (minimum) Solar charge controller

A 12 volt blade fuse box & fuses

Cables & wires

Bus bars

Wire connectors

12 volt battery

Terminal blocks

12 volt power sockets

12 volt bulbs & bulb holders

Don't worry if you don't know what these items are as they will be covered in detail throughout the book.

It is possible to buy package deals that have the solar panel, the charge controller and the necessary cables all together for a reasonable price on eBay so it's worth keeping your eyes peeled for any these sensibly priced bargains.

Chapter 3

Buying the right battery

Twelve volt batteries come in different shapes and sizes with varying amounts of terminals on top of them for connecting cables onto. There are two types of 12 volt batteries that people will tend to use on 12 volt solar systems, the car battery and the leisure type battery.

A car battery is constructed using thin lead sheets so that it can give out massive bursts of power in order to start a car engine. This is what a car battery is designed to do best. It can give out small amounts of power for feeding 12 volt equipment but this is not what it was designed to do, so it is not the ideal battery for use on a 12 volt power system and its lifespan may not be that great because of this.

A 12 volt leisure battery (which can also be referred to as a deep cycle battery) however was never designed to give out huge amounts of power in one go. It has thick chunky lead plates inside it that make it more suited for giving out smaller amounts of power over very long periods of time. This makes a leisure battery ideal for powering your TV's and USB devices on a 12 volt solar power system. You will find that the leisure battery has four terminals on the top of it so that you can connect extra cables to it securely in order to power your devices properly.

This book is primarily written for those who only need reasonably small amounts of off grid power. If this is the case you will probably get away with only using one battery for your power needs and this is what I advise you to buy at first. It is a very simple matter to purchase another battery if you find that you require more power storage capability later on.

Small warning: When buying extra new batteries ensure that their power & storage capabilities match your old ones. The Voltage and Amp hour rating should be the same as the battery or batteries that you already have. This ensures that they all match up in their capabilities and prevents any problems that you may have with charging them.

Connecting batteries together

If you find that you do need more than one battery then you will have to put some thought into how you connect these extra batteries together. You have a choice of connecting them either in a Series or Parallel configuration.

Series configuration

If you connect the batteries in series then the voltage on your 12 volt system will change and it will no longer be a 12 volt setup. Two batteries in series would become 24 volts, three would become 36 volt and it would keep going up every time you added another battery. You may not be familiar with series & parallel connections so here is a diagram of a series connected set of batteries.

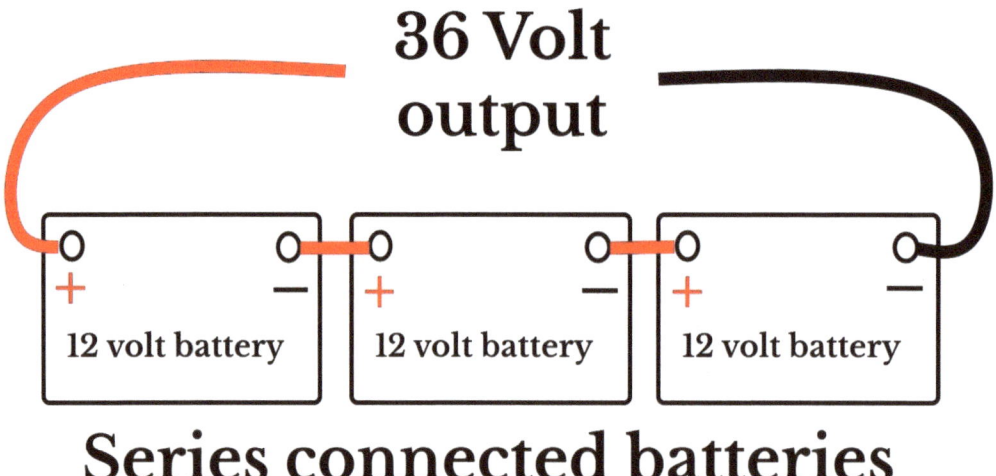

Series connected batteries

You can see from the diagram that as you connect the batteries together in series that the voltage increases by 12 volts for each battery connected which is what you don't want.

Parallel configuration

Connecting the batteries together in parallel will give you the ability to store more power whilst keeping the voltage at 12 volts.

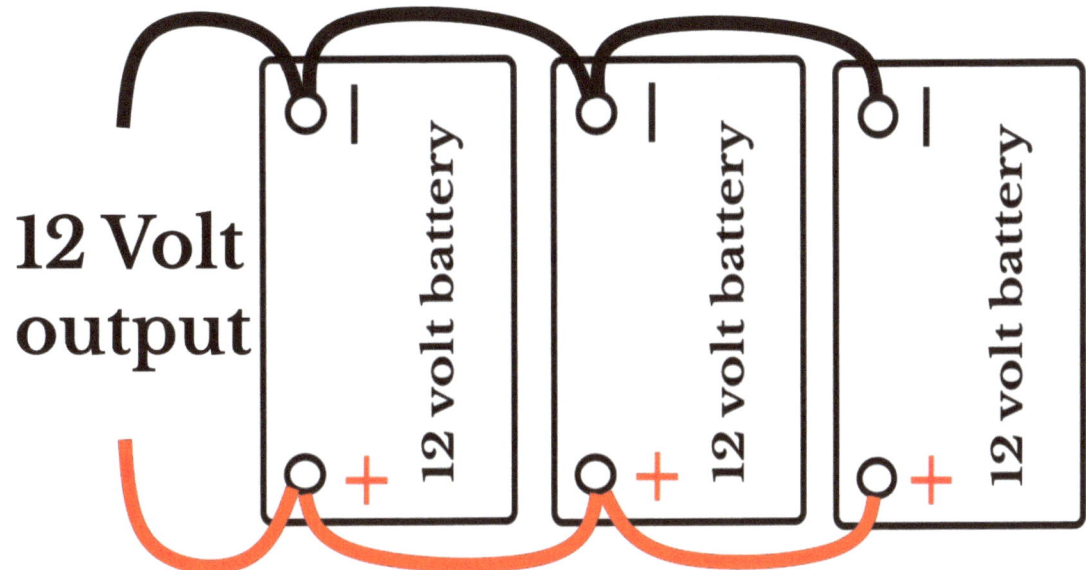

This diagram of a parallel connected set of batteries is a much better setup for your needs. The voltage stays at 12 volts and your power storage capacity is tripled in this particular setup. All this extra storage can make the difference between running out of power on a regular basis and never having to worry about running out ever again.

Hopefully your power needs may not be overly great and one battery may serve all your requirements. If however you find that your power supplies are running low then you can always buy another one or two batteries and link them up in parallel at a later date.

How much power can a battery hold?

Although batteries tend to be similar sizes externally, internally they are capable of storing different amounts of power. This ability to store power is called an Amp hour rating, which is often just referred to as the AH rating. The bigger the AH rating the more power that particular battery can hold.

If you have a 100 AH battery and you are powering a device that needs 10 amps to run it, then basic division tells you that your battery can run this device for around ten hours before it flattens the battery. In reality you don't get quite this long but it is close enough to give you a rough estimation of your power needs and what your battery can provide you with.

All the devices that you wish to power use something called Watts or Amps to power them. Any device that you intend to power should have its power requirements written on it somewhere (usually by the power lead). It is easier to understand how much power a device will demand from your battery and how long your battery can keep feeding this device with power if the devices power demands are marked in amps, it just makes life simpler.

If the device that you wish to power is rated in watts then fear not, all you have to do in order to get its amp rating is to simply divide the voltage that you are powering it with by the watts that the device uses. This voltage in your case will always be 12 volts because you are using a 12 volt system.

This formula can be adapted to work with any voltage that you are powering. Even devices using 24 volt, 110 volts and 220 volts can have their amp rating worked out using this method. In your case you are working with 12 volts so all you do is simply divide this voltage by the watts and you will have the amp rating of your device.

If you are using a device that states that it uses 10 watts of power, then dividing the 12 volts by the 10 watts gives you a device amp rating of 1.2 amps.

$12 \div 10 = 1.2$

The 100 amp battery mentioned earlier should be able to power this device for about 83 hours (100 ÷ 1.2 = 83.333333) if this is all that it has to power. By working out the power demands of all your devices you can get an idea of how much power you will need to store to meet your needs. You don't have to do this but it does make working out power usage and battery storage capabilities a little bit easier.

Measuring how much power is left in your battery

Once you start using power from your battery your first concern will most likely be how much power you have used and how much is left for you to use. This is an easy thing to discover by using a simple voltmeter that will give a digital display of how much power is inside your battery.

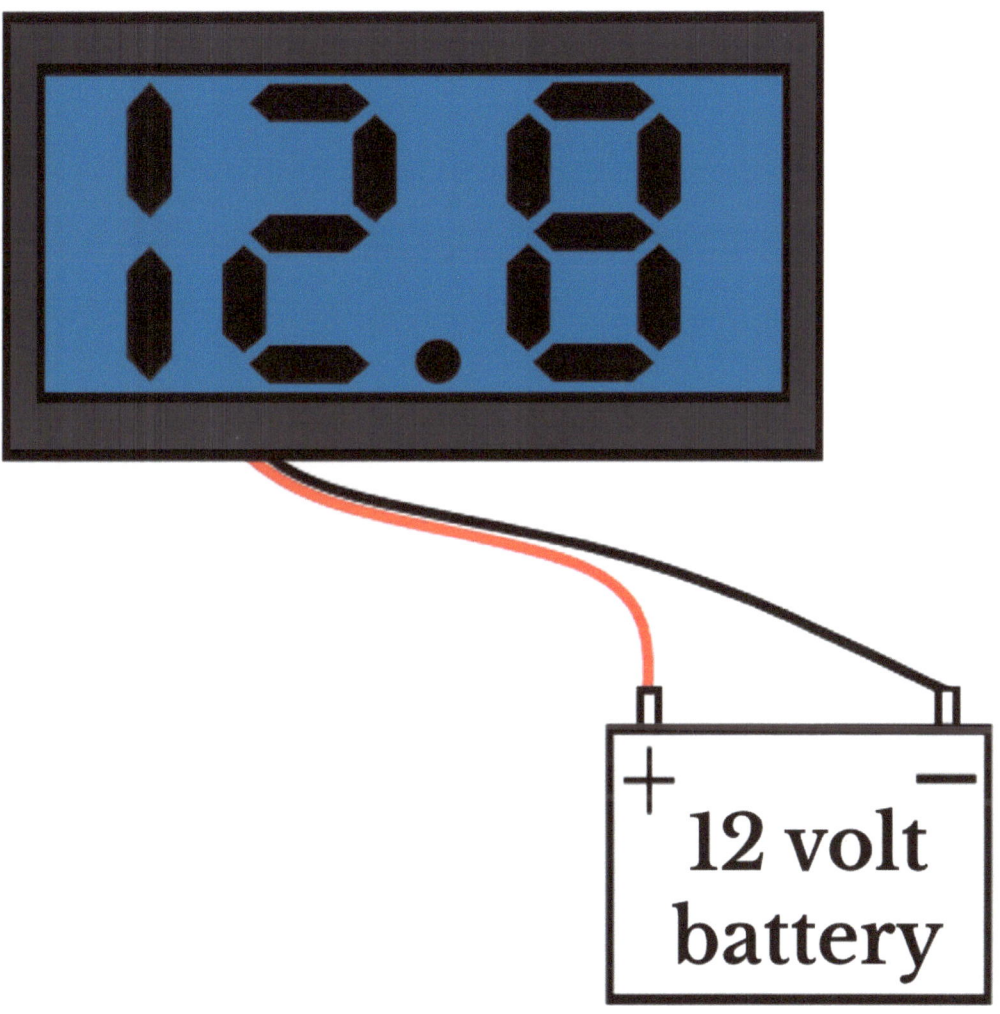

A voltmeter is a simple and cheap device that is simply connected to your battery and it measures the voltage that is in the battery. The more power you use the lower the voltage goes and the more power you place into your battery the higher the voltage will go.

Although it is a 12 volt battery the reading on a fully charged battery will be around 12.8 volts (almost 13 volts), this is a reading that you like to see. When you have used around half the batteries power supply the display should read around 12.4 volts. A reading of 12 volts or below it means that your battery has been drained and needs recharging straight away.

Battery power level warning: You should never allow the voltage of your battery to drop below 12.4 volts because your battery doesn't like dropping below the halfway point and it can actually shorten your battery life if you do this. This means that the 100 AH battery mentioned earlier is in reality only a 50 AH battery if you want to make it last as long as possible. Keeping your battery topped up by using a solar panel means that hopefully your battery will always have the maximum amount of charge in it that is possible. This will keep your battery healthy and make it last longer.

Chapter 4

Choosing a solar panel

Most people who use a 12 volt system have at least one solar panel to charge up their batteries. Installing and using a solar panel is not difficult to do even if you have had no experience with these in the past.

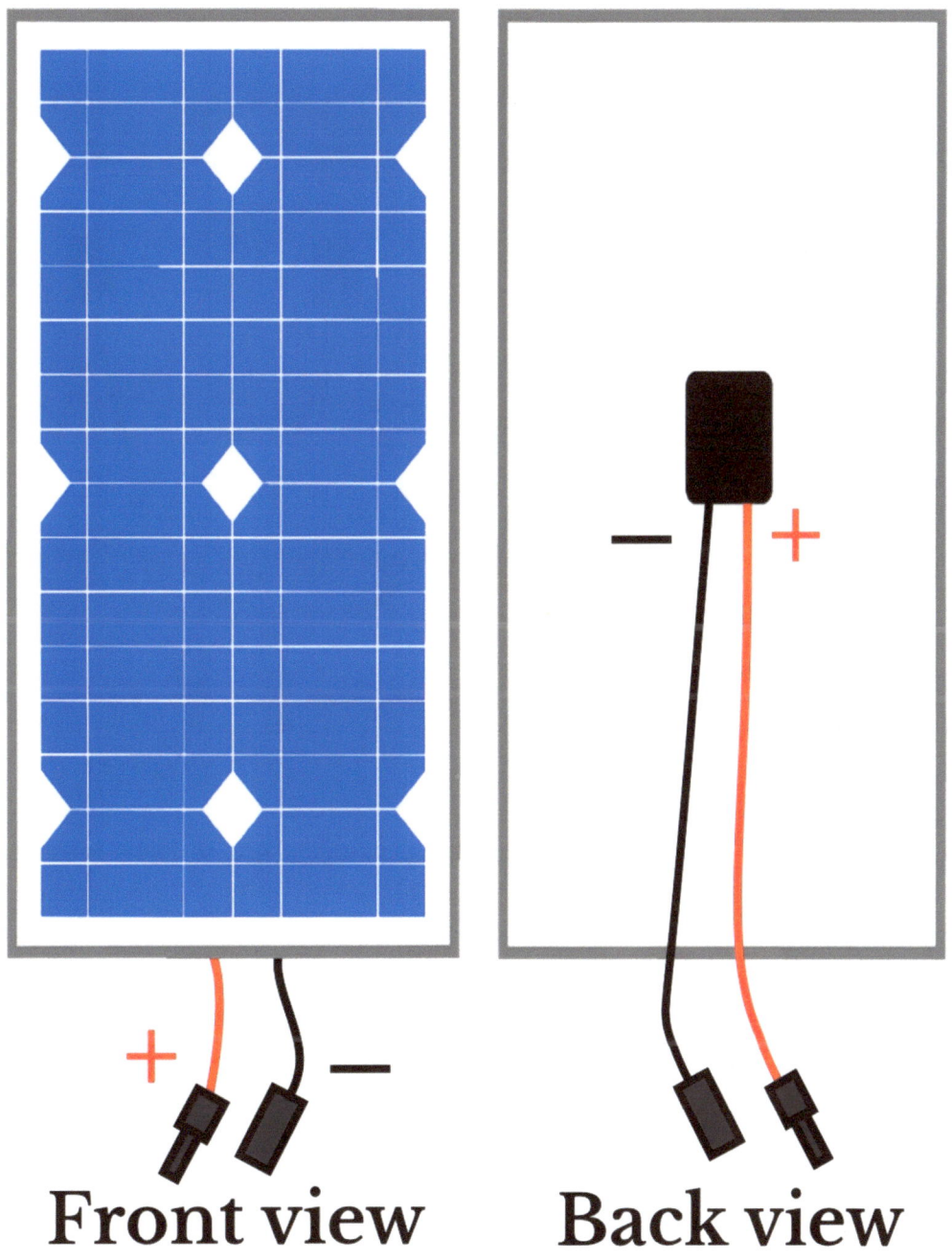

In reality both wires on the back of a solar panel are black in colour. The wires in the picture are a different colours to emphasise the fact that there is a difference and you must pay attention to which wire is marked with a "+" sign for the positive connection and a "-" sign for the negative connection.

A solar panel can be used as one individual panel or they can be linked together in parallel. Linking them in parallel (not series) keeps the voltage the same but increases the amount of amps that the solar panels put into the batteries for you.

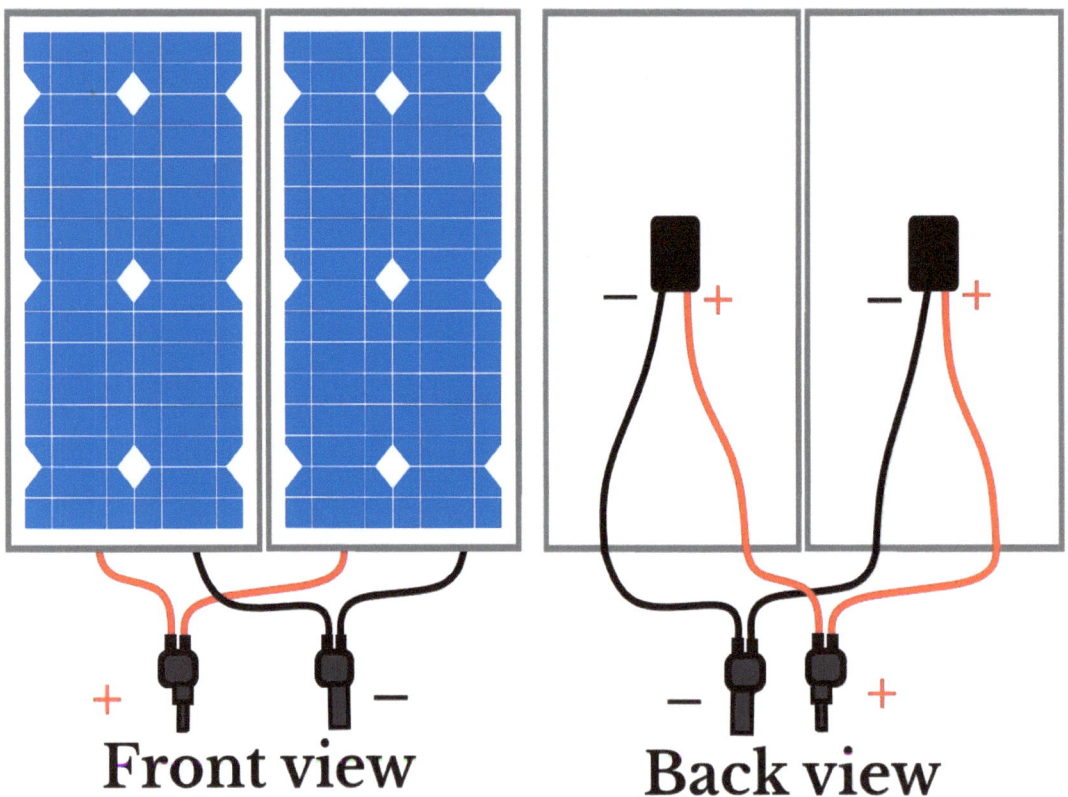

Solar panels come with male & female fittings on the ends of their wires called MC4 connections. In order to connect multiple panels you simply connect the panels together via a connector called an MC4 branch connector or Y connector.

Whether you have one panel or multiple panels you should only ever end up with two wires that connect to the solar charger, one positive wire and one negative wire.

You use an extension wire that runs from your solar panels and runs down to the inside of your shed. On one end of these extension wires will be a MC4 connector to attach to the plug that comes from the solar panel and on the other end it will just be a bare wire which will be attached to a solar charge controller inside the shed.

For use in a shed, boat or van I would recommend that you start with a 100 amp solar panel minimum in order to provide enough power to charge up your batteries and power your devices in the beginning. If you find that you need more power then it's a simple matter to buy a second or a third solar panel of equal power and link them all together in parallel.

Once your solar panels are all setup on the roof you should have two wires that come down and into your shed, boat etc. The positive and negative wires that you have are simply connected to a power management device called a solar charge controller.

It comes as a surprise to some people that a solar panel doesn't generate 12 volts of power it, and that it actually generates around 18 volts of power. The reason it does this is because batteries actually need a charging source of over 14 volts in order to push the raw power back into the battery itself. With the voltage being at around 18 volts the solar panel easily generates enough power to charge the battery itself. Unfortunately if you connected the solar panel straight to the battery itself then all this raw power would overwhelm the battery and damage it. It is for this reason that a control device must be used to handle the current from the solar panel and regulate how much of this power is fed into the battery itself. This control device is called a solar charge controller.

Solar charge controllers

A solar charge controller takes all the power from the solar panels and feeds the correct amount of this power to the battery in order to charge it up. When the battery is full it stops the charging process until the battery needs more power again. This keeps the battery charged up to its limit whenever there is enough sun to charge it up, but prevents it from over charging the battery which would damage it.

The solar charge controller is only a very small device but it is shown as much larger in this diagram for demonstration purposes.

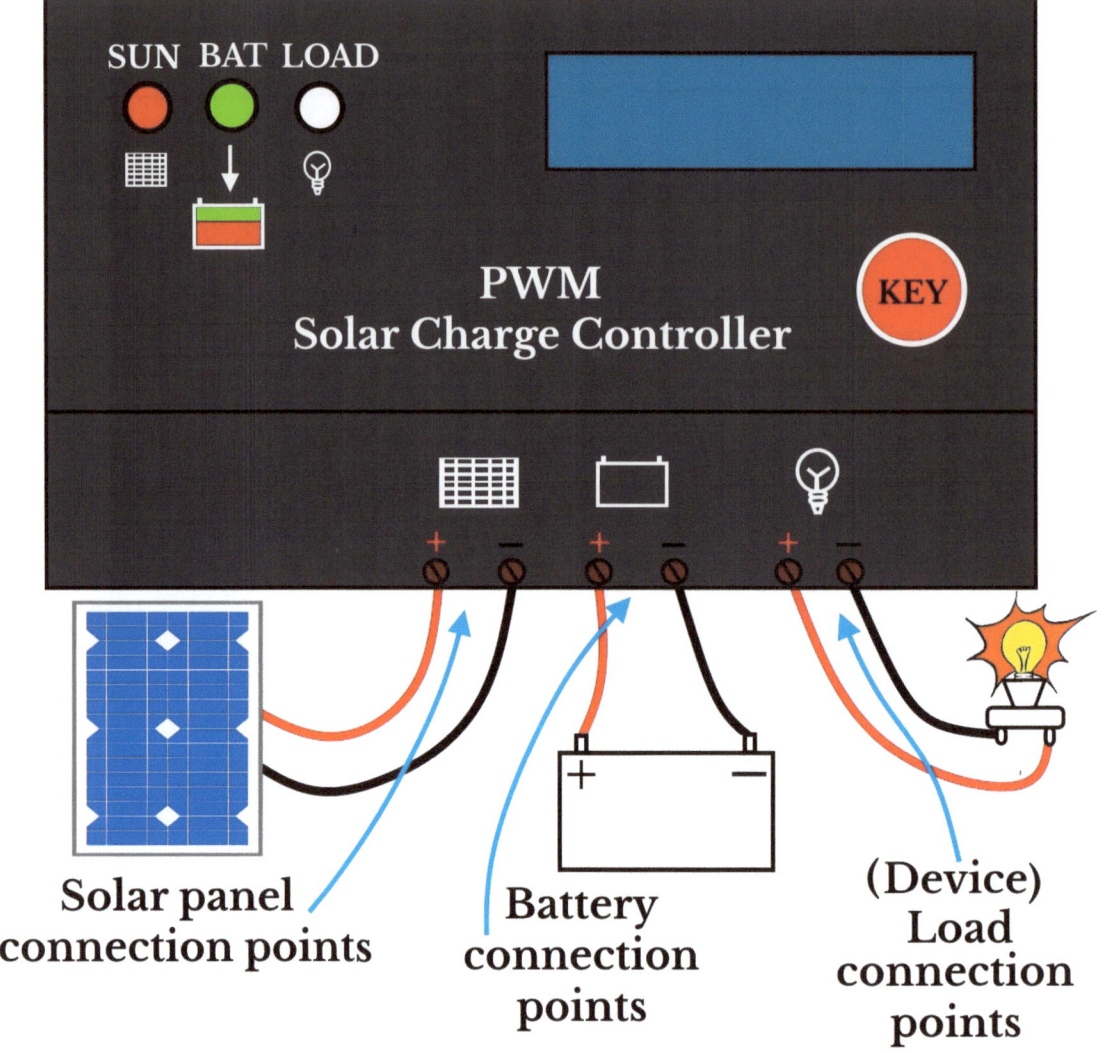

Solar panel connection points

Battery connection points

(Device) Load connection points

It is actually very easy to wire up a solar charge controller as you can see from the diagram. The solar panels are connected to the first set of screws. The battery connects to the second set of screws so that it can be charged by the charge controller. And you can even connect a device or two (power socket, lights etc) to the load section (light bulb picture) via the other set of screws. If all you want out of this system is to power some lights or a power socket for phone charging etc, then you can usually connect it to the load section and your system is finished. So long as your charge controller is amp rated above your power needs then you can use the load option.

There can be a digital display to show the voltage or settings on the charge controller but this is not always the case (especially with the cheaper ones). There are usually a few lights that flash to show you that it is receiving power from the solar panel, the status of the battery levels and if power is being drawn from the load section.

It is a simple case of connecting your wires into the positive & negative holes on the solar charge controller and securing them in place with the screws, it really is that simple. You do need to ensure that you use the correct thickness of solar panel wire for the connection between the panel & the charger, keeping in mind that the solar panel operates at around 18 volts of electricity rather than 12 volts that the rest of your system runs on. If you buy a solar panel kit then the extension lead that it comes with will usually be thick enough for at least two panels of the same voltage, especially if you keep this lead fairly short.

There are two main types of solar charge controllers that you can buy called a **PWM** (Pulse Width Modulation) controller or the **MPPT** (Maximum Power Point Tracking) controller.

The PWM controller is cheap to buy & easy to install. It is a very simple device to use and understand. And whilst it may not be as efficient as the MPPT controller it is much more durable and simpler to use.

The MPPT controller is very expensive to buy and they are a little sensitive to vibration or being knocked so they may not be best suited for vans etc. They are more efficient at converting power than the cheaper PWM controller but they are about ten times more expensive to buy. They work better on large arrays of solar panels than they would on only one or two panels so you may be better off sticking with the PWM controller.

For the purposes being explained in this book the PWM charge controller is the cheaper & most preferred option for a solar setup of this size.

Load option warning: you may decide to use the load feature on the charge controller to provide the power for your devices. If you decide to do this then it is sensible to get the highest rated solar charge controller that you can afford. This will ensure that it has a high enough amp rating for your needs. The load function can only provide the same amount of power that it is rated for. So if you intend powering a 10 amp socket and some lights on the load option then a 10 amp rated solar charge controller won't do it. A 30 amp (or larger) solar charge controller would allow you to use two 10 amp sockets and have some low power lights switched on at the same time suiting your needs much better.

Important safety warning: Your solar panel will push electrical power out any time that it is in the sun, so if you are connecting it to the charge controller or working on the system then you should cover the panel up to stop the flow of electricity while this is done. Another alternative is to place a suitably rated switch on the positive cable so that it can simply be disconnected with the flick of a switch, making it safe to work on. This makes it safer and much more convenient.

It may be easier for you to buy a solar panel kit that has the solar panel, an extension cable and a solar charge controller bundled with it. This way the bulk of your system is delivered to you in one go. These are available from eBay & Amazon and will be delivered straight to your front door.

Chapter 5

Creating a basic safe circuit

There are three basic designs of circuits that you are able to use on your 12 volt solar setup, these circuits are called the basic circuit, the series circuit & the parallel circuit. In this chapter we shall only cover the basic circuit because this is the one that you are recommended to use at the moment. In the last chapter series and parallel circuits will be covered so that you understand them and you will also have the option to use them if you so wish to do so, but for now we shall just learn the basic circuit design because of its simplicity and safety.

The basic circuit

The basic circuit is basically the safest circuit in the world. It has only one device to power which makes selecting a suitable fuse for the device extremely easy. A basic circuit consists of one single wire which goes to one single device and then returns to the battery. This single device can be a light bulb or a power socket or any other device. Only one device at a time is ever connected to this circuit. This does limit its usefulness to just one device at a time, but it also means that you have no need to work out the amp ratings of a device on this circuit because it should be written on the device itself. Selecting a fuse for this type of circuit is child's play because the amp rating for the fuse is also the same as the amp rating that is written on the device that you are going to connect to this circuit.

For those of you who have no understanding of an electrical circuit I shall attempt to talk you through the basics.

A battery has a positive (+) & a negative (-) connections points on them called terminals, as do the devices that you wish to power. Electricity comes out of the positive terminal on the battery (+) and this is where we attach the positive (red) wire to. The other end of this positive wire is attached to the positive terminal on the device (light bulb etc) that you wish to power. This allows the power to flow from the positive terminal up the positive wire and force its way through the device. It then comes out on the other side of the device through the devices negative terminal. Attached to this negative terminal on the device is the negative wire (black). This negative wire is attached to the negative terminal on the battery.

If this all sounds a bit complicated then take a look at the next diagram which will hopefully show you how simple this process is.

Basic circuit diagram

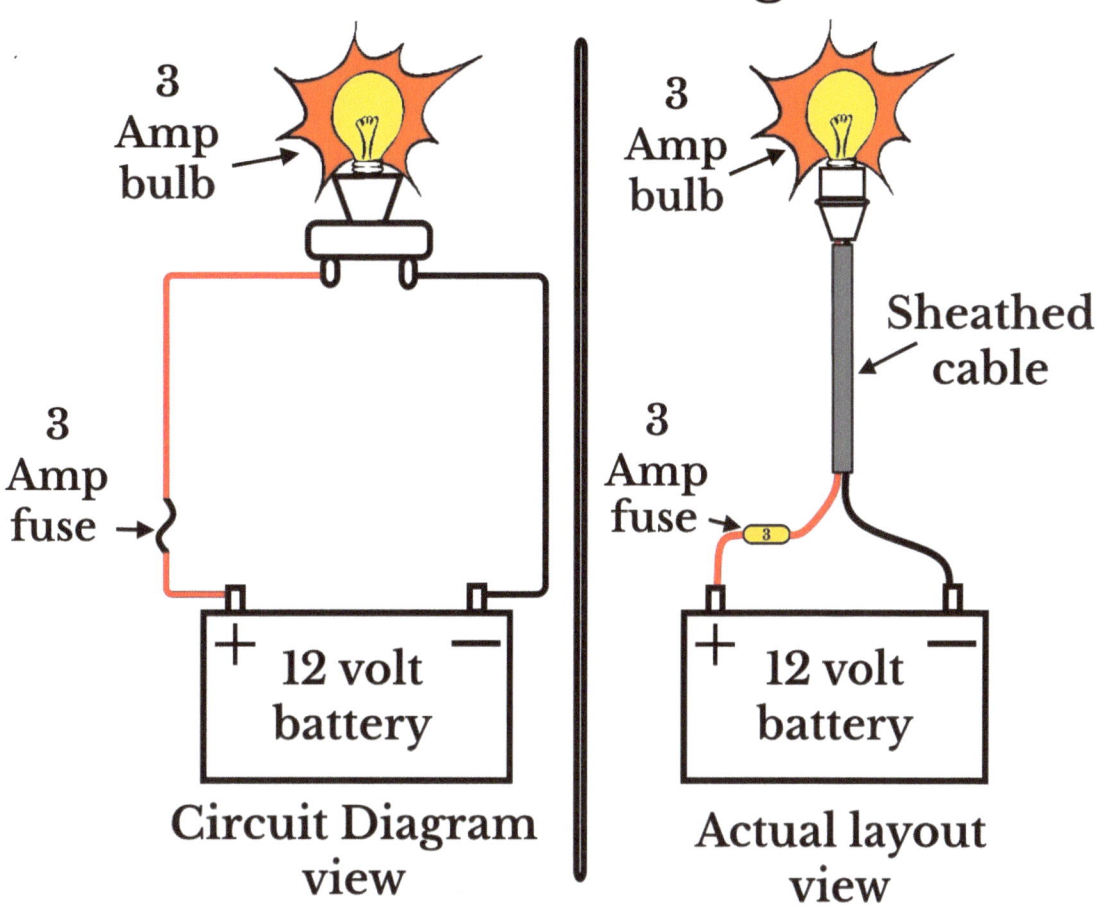

Circuit Diagram view

Actual layout view

This simple loop of wire allows power from the battery to travel around it pushing power through the device as it does so, which makes the device work. If this wire is disconnected from a terminal then the power will stop flowing and the device will stop working.

The circuit on the left is a schematic circuit type diagram, you will see more of these later in the book. I have tried to show you how the circuit would actually be constructed in reality by placing a more realistic setup on the right hand side of the diagram.

As you progress through this book you will see how to connect multiple circuits in order to fulfil all your power requirements. You will be using this basic circuit and you will find it amazingly simple to put together your system using this simple circuit. Once you have built your system using this method then you can experiment with using series circuits and parallel circuits but for now concentrate on these and you will be shocked with how simple they are to use and how quickly you will construct a fully functional system.

Chapter 6

Connecting things together

Now that you know how a basic circuit works you need to know how to attach your devices to your circuit safely and easily. This is achieved using wire connectors and terminal blocks, which we will cover next.

Terminal connection blocks

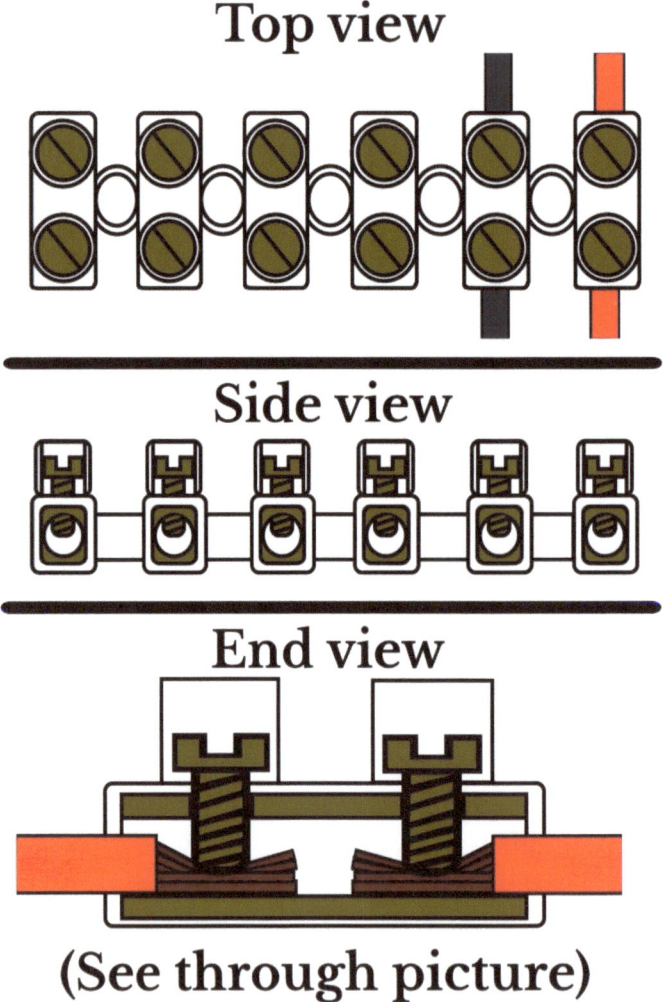

To connect two wires together or connect wires to a device we can use something called a terminal block. This plastic block has a metal centre which has a hole drilled through it into which you place the ends of the wires that you wish to join together. There are two screw heads that you can tighten which will hold these two wires in place. Doing this connects these two wires together so that the power can flow through them both, in effect these two wires are now acting like one solid wire.

These blocks are ideal for connecting your wires to a device so that you can power it. Your power sockets, lights, volt meters etc, can easily be attached to a circuit using these.

The terminal blocks come in strips so that you are able to make multiple connections or they can be cut down to just the right amount of connectors to suit your needs. This gives you lots of options when you are building your circuits.

Using the terminal block

This is a simple diagram of how to use the terminal block to connect your wire to your devices that only have wires on them and don't utilise metal connector pins.

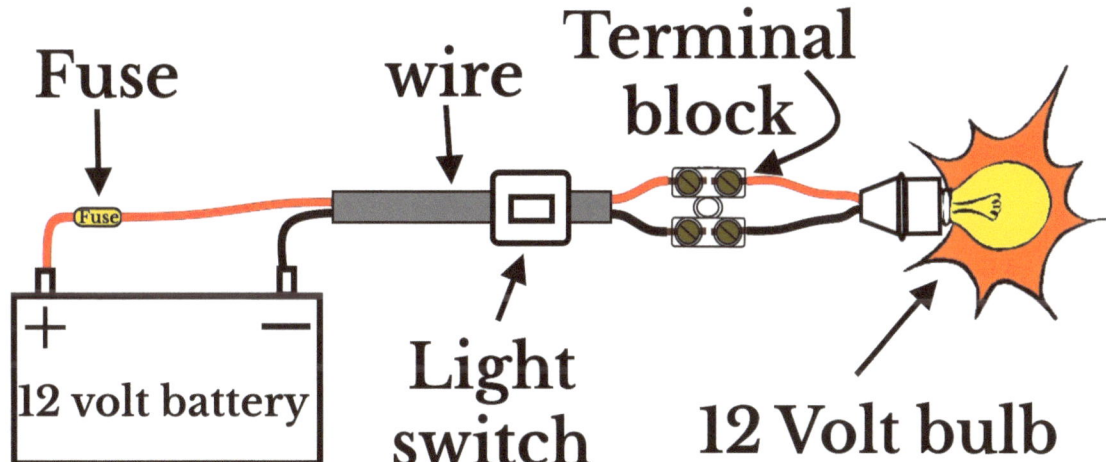

These terminal blocks are cheap to buy and are available from most hardware shops and even some supermarkets. They are amp rated so you should go for the amperage that suits the circuit that you have built, although it is always safer to go for a higher amp rated terminal block than is actually needed to give your circuit a higher safety rating. They are very simple to use so no previous experience is needed when using these.

Spade end connectors

Another form of wire connector that is frequently used is the spade end connector which will prove invaluable to you when you are building your system.

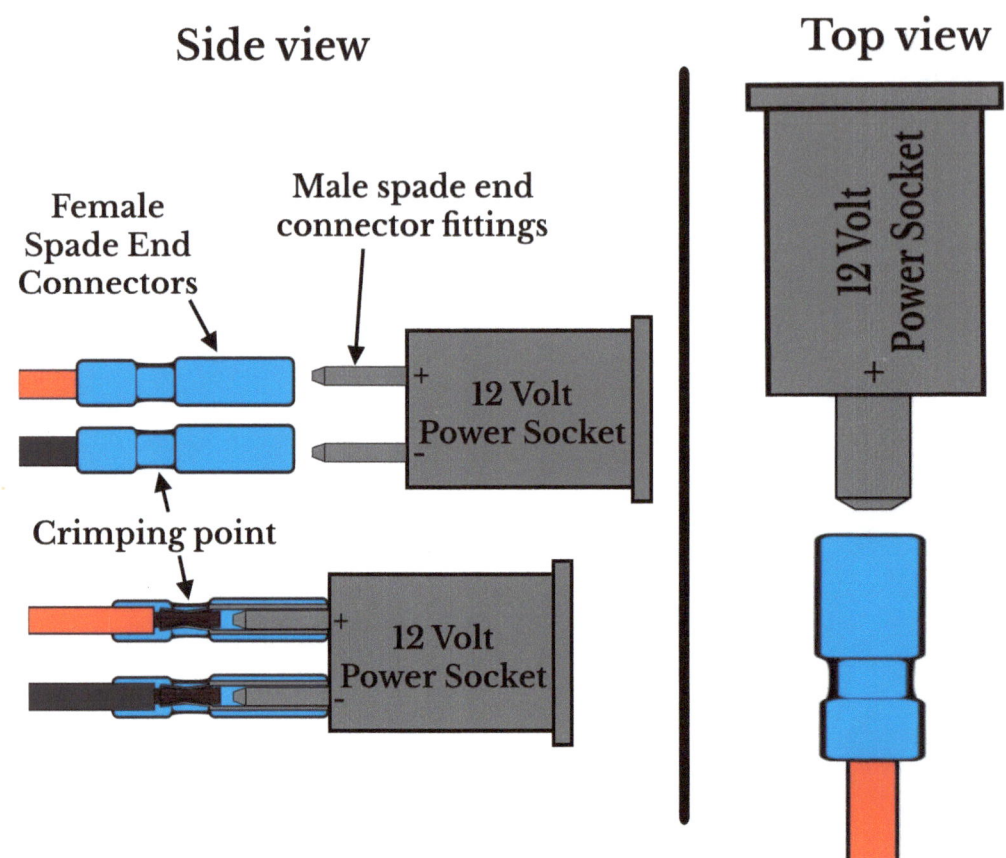

You will see the spade end connectors inside this book on the ends of the wires in the diagrams. Because there are male & female spade end connectors it means that wires can actually be connected together using just a male & female connector.

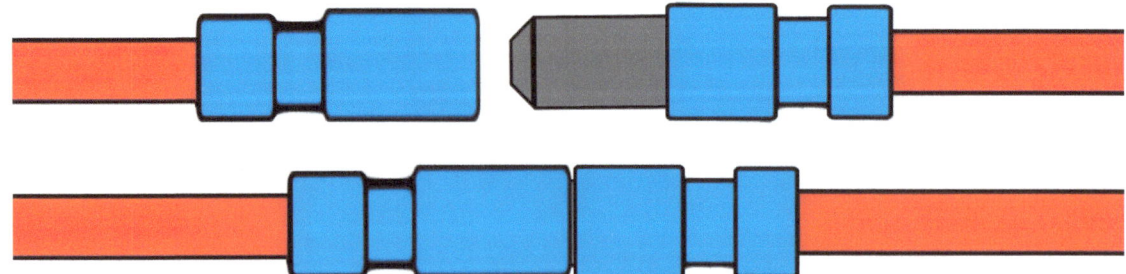

It is fairly rare that you would need to join two wires this way as these connectors are more often used for attaching your wires to various devices, but it can be required on occasion.

These connectors fit over the ends of the wires and are crimped into place using a set of wire crimpers. Crimping a connector means to crush the body of the connector over the bare ends of the wire that you are attaching it to. This should give you a strong secure join if you have crimped it properly.

These connectors allow you to attach your wire to any device that has a male spade end fitting attached to it. Fuse boxes, power sockets and many more devices make use of these connectors so it is well worth picking a few spade end connectors up for your parts kit.

Ring end connectors

Ring end connectors or round end connectors as they are sometimes called are very useful for attaching wires securely to bus bars and other devices.

They have a circular connection point that has a hole through it so that they can be slid over a bolt and secured with a nut (or simply screwed into place). These connectors give a very secure and safe connection point that won't come loose unless they are unbolted or unscrewed.

Sizing the connectors

The spade & ring connectors are colour coded for the size of wire that they are made for. Generally speaking the blue connectors are the ones that you will use the most but as soon as you start using thicker wire or you are crimping two wires into one connector then you will need a larger size connector. Below is a rough guide to help you select the right colour connector to suit the size of your selected wire. You will find that these three colours will generally cover all your possible needs in wire sizes.

Sizes chart

Yellow   =   10 to 12 AWG

Blue   =   14 to16 AWG

Red   =   16 to 22 AWG

You will see these connectors in use in the next chapter when the fuse box is setup & the bus bars are created.

Chapter 7

Fuse boxes and keeping it all safe

Fuses are designed to keep a circuit safe. The fuse is made of thin metal that is designed to melt when the current goes above a certain amount of amps. This way if more current (amps) flows through the wire or the device than it was designed to handle then this thin piece of wire will melt and prevent the electric current from flowing anymore. It will stay like this until a new fuse is placed into the fuse holder.

There are a few different types of fuses for various jobs and applications. For the 12 volt circuits that you shall be making we shall be using your common 12 volt Blade type fuses that are used in vehicles.

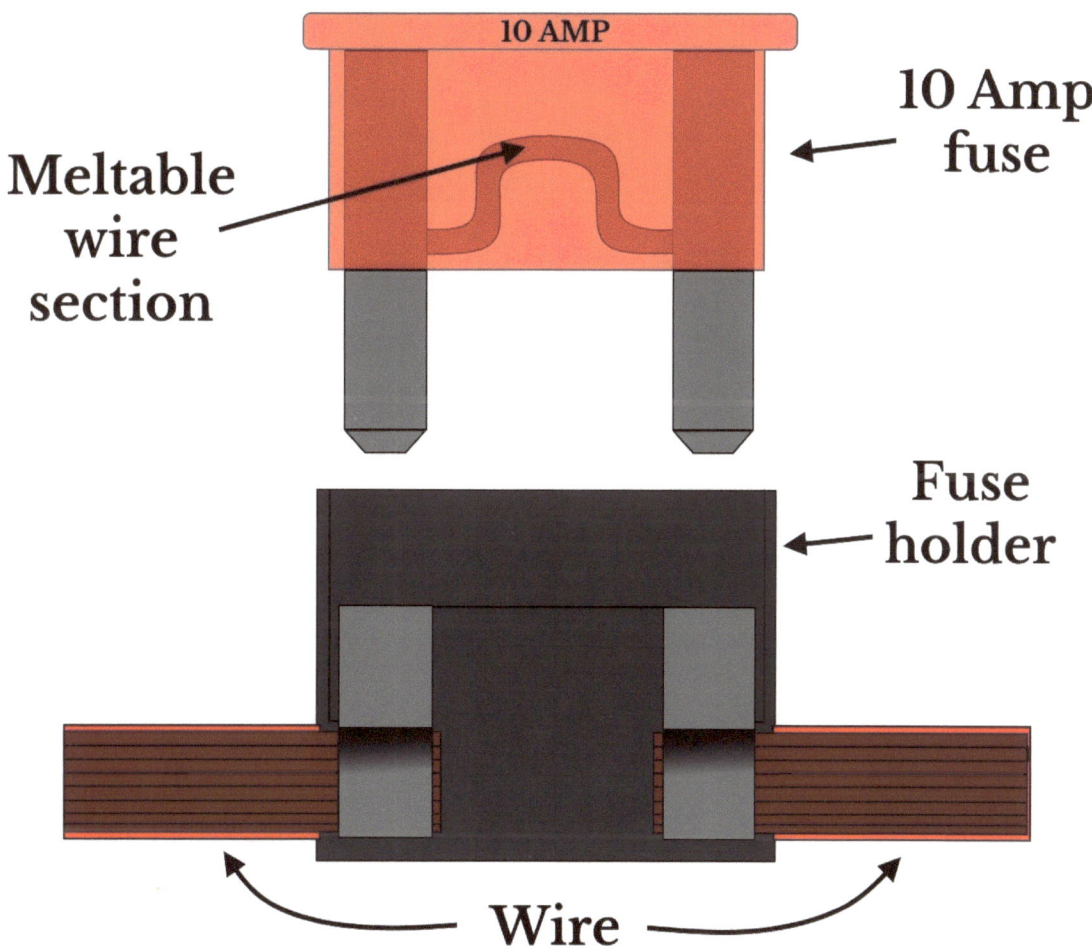

These fuses are constructed of a rigid see through plastic that enables you to see the thin metal wire that is designed to melt in the case of an overload of current. This way you can see instantly if the fuse has blown or not. If the thin metal wire is intact then the fuse is usually good but if the wire has melted away in the centre then there was an overload of power and this fuse is of no use anymore and it needs to be replaced.

The wire that the power travels along is broken and will not allow power to travel along it unless there is a working fuse in the fuse holder itself. The power travels down the first piece of wire and then through the fuse to get to the other piece of wire so that it can continue its journey to your device and then back to the battery.

You can use an individual fuse holder like the one in the last diagram to protect your circuit or you can use a fuse box. A fuse box is a small plastic enclosure that holds multiple fuses and each of these individual fuses is attached to an individual circuit. If you buy the right size fuse box then you should be able to power every circuit on your 12 volt system from this one fuse box.

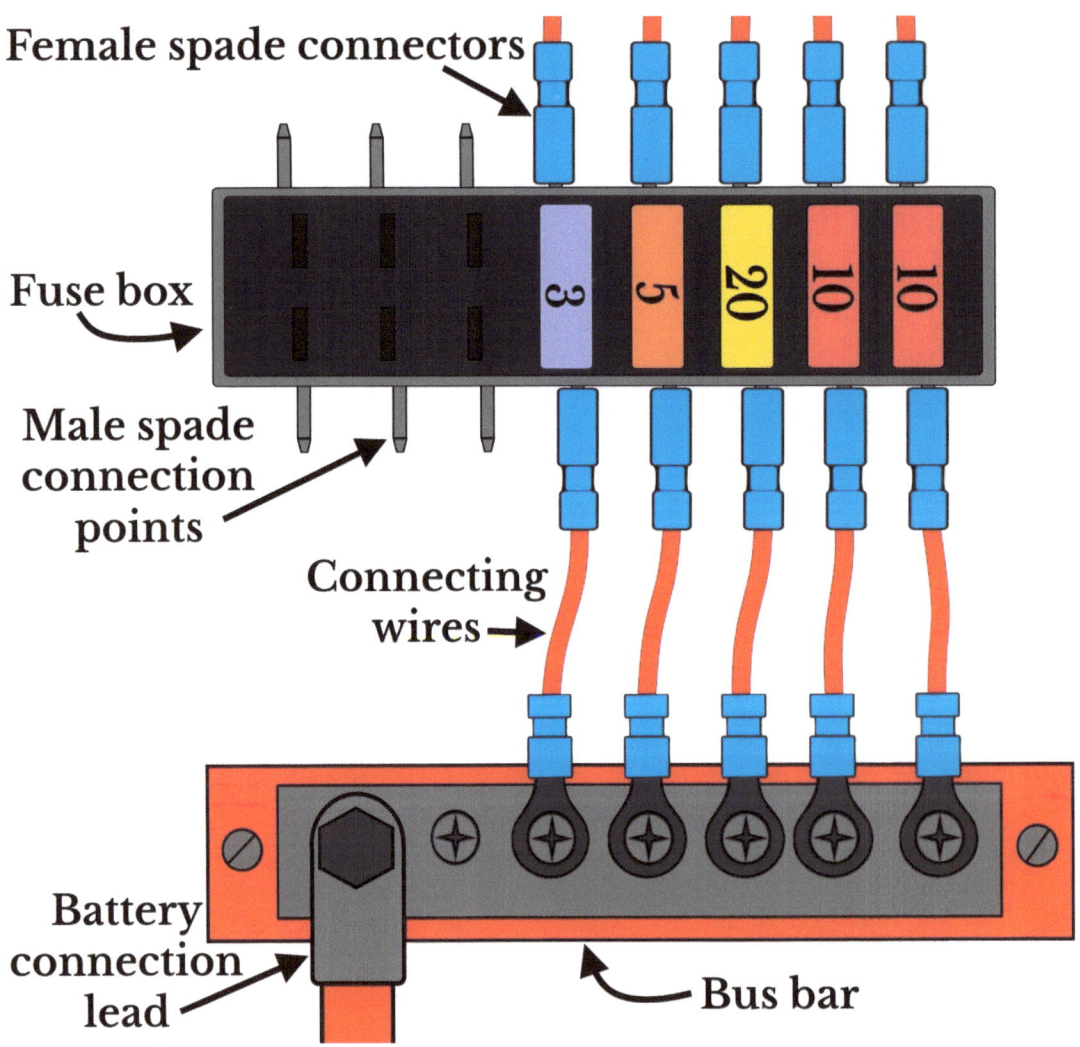

The current for each individual circuit on your system flows in and out of the fuse box. If the fuse is removed then no current will flow and that circuit will not work until a new fuse is placed into it again. Using a fuse box rather than individual fuse holders tidies your system up and makes it look more professional and means that the whole system can be managed from one location.

There are various colours of these fuses with each colour having a different amp rating. The different colouring system allows you to see at a glance what size fuse you have in the fuse holder from a distance. It does also have the amp rating written on the top of the fuse itself so you should never have a valid reason for not selecting the right fuse for the device that you are protecting with the fuse.

Devices have an amp rating written on them because they need a certain amount of amps in order to function properly. If this device at some point develops a fault or a short circuit then it may try to draw more power than it can safely handle. This can cause the device to overheat and possibly burst into flames. This sudden surge of power is what a fuse is designed to prevent. If a device is rated for 3 amps of power it should never actually need more than this to function properly (except for momentary power spikes). This is the given figure on the device and you know that if your fuse on the circuit is also rated at 3 amps then there should never be a problem with running this device on this circuit. It is possible that some devices need this full amount of amps and possibly a smidgen more when they spike at start up or when performing certain tasks. If this is the case then a 5 amp fuse may be a better choice for this device rather than allowing it to constantly pop fuses, but generally speaking it is always better to go for the fuse that matches the amp rating of the device. Don't forget if the device only has a watt rating then simply divide the 12 volt rating of your system by the watt rating to get the amp rating of this device.

If the device does develop a fault and demand more power than it is rated for then the fuse should melt before either the wire or the device have a chance to and stop the flow of power through the wire itself making it safe.

If your device needs a certain amount of amps that doesn't fit the fuse that you have then don't put a smaller fuse into the circuit i.e. if a device requires 5.5 amps then a 5 amp fuse won't do the job because the fuse will most likely keep popping. A 7.5 amp fuse would be the next closest fuse that you are able to get so that would be the best one to use in this situation. The fuse must be equal to or very slightly greater than the required amps otherwise it will continually blow the fuse.

The bus bar

A bus bar like the one in the previous diagram is usually made of a solid metal strip that can conduct a lot of electricity along its length. Onto this strip you fasten a ring connector for each individual circuit on your 12 volt power system. Each circuit takes the power it needs from the bus bar and moves it along the circuit to power the device that you have wired onto that circuit.

Although you can buy these professional type bus bars online, I prefer to make my own when making very small systems that I would place into a shed etc. A reasonably large nut and bolt can easily be used and because they are made of solid steel they should be more than adequate for handling the small amount of power needed within a small setup like this. They cost pennies to buy and reduce the cost of the overall setup.

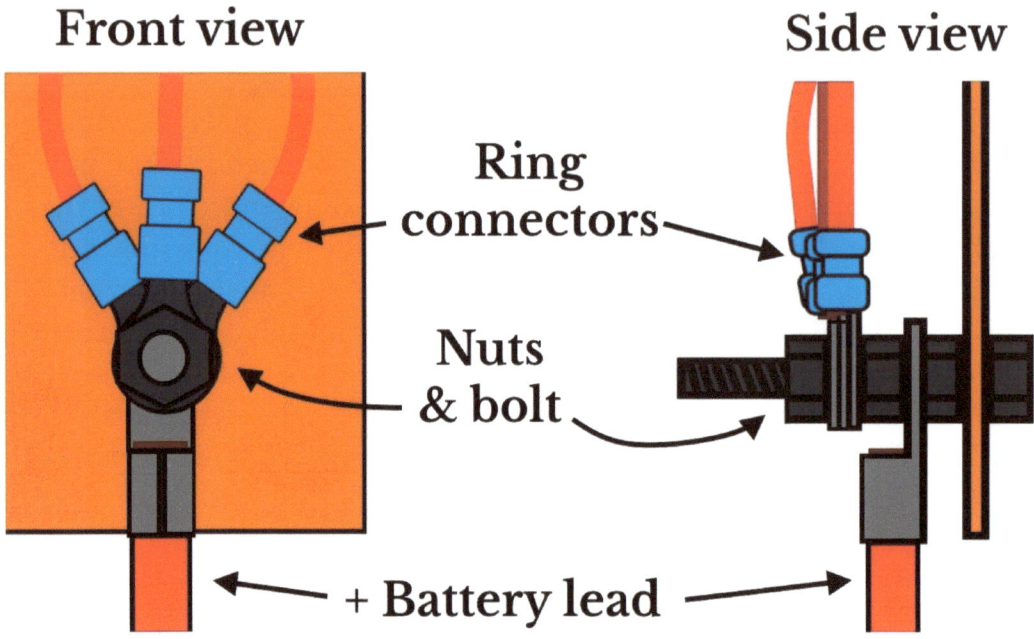

This improvised bus bar can be created and connected up in less than a minute so it is an ideal addition to your system.

Chapter 8

The switch

You will need switches to turn your devices on and off, so learning how to connect these switches is essential for you. So here are the basics that are involved with fitting these.

There are two types of 12 volt switches for you to choose from, the two pin and the three pin switch.

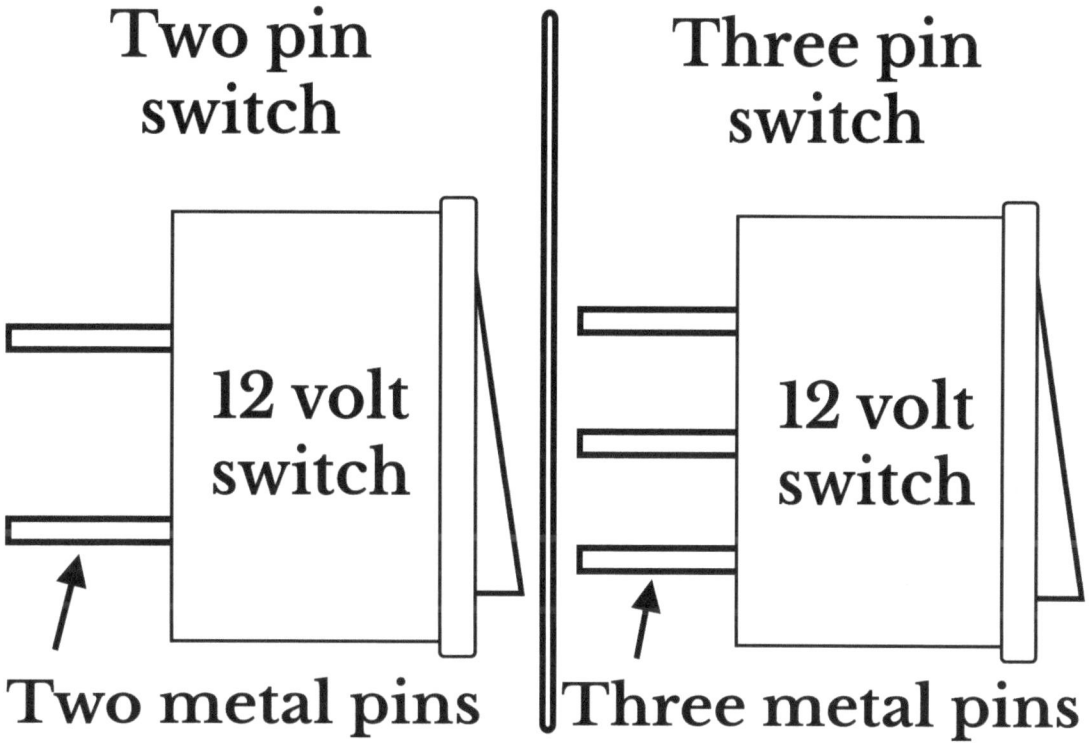

A switch is basically just a break in the wire. The switch itself is either in an open position which prevents the current from flowing through it, or it is in the closed position which allows the power from the battery to flow through the switch and into the other wire which leads to the device.

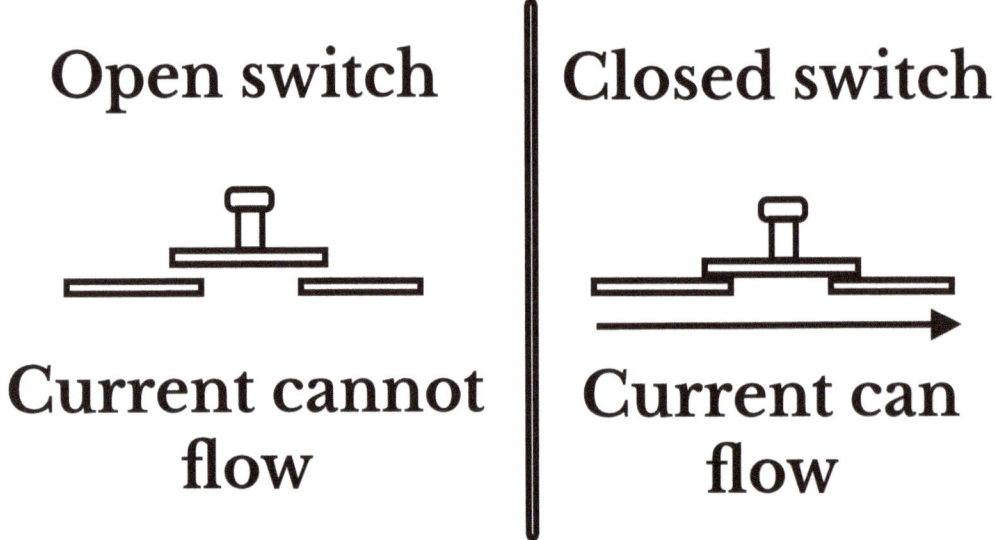

When the switch is in the closed position power can flow to the device that you are powering. The power will go through the device itself and then return to the battery via the black negative lead.

The basic two pin switch

Wiring the two pin switch is fairly easy as you can see from the following diagram.

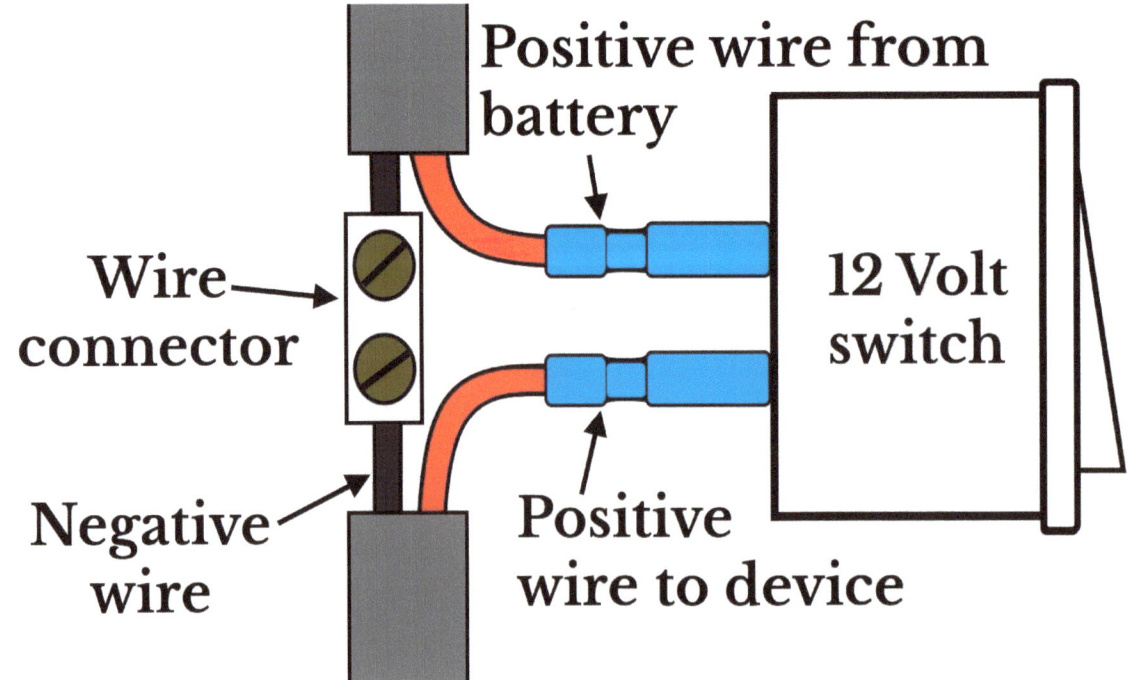

The positive wire from the battery is attached onto one pin. This allows the power from the battery to travel from the battery and into the switch. It then travels through the switch and out of the other pin so that it can power the device. It then travels back to the negative terminal connection on the battery. When the switch is in the closed position the power will continue onto the device, but if the switch is in the open position no power can flow.

The three pin switch

Wiring the three pin switch is a little bit more complicated but it is easy to do once you understand the basic setup.

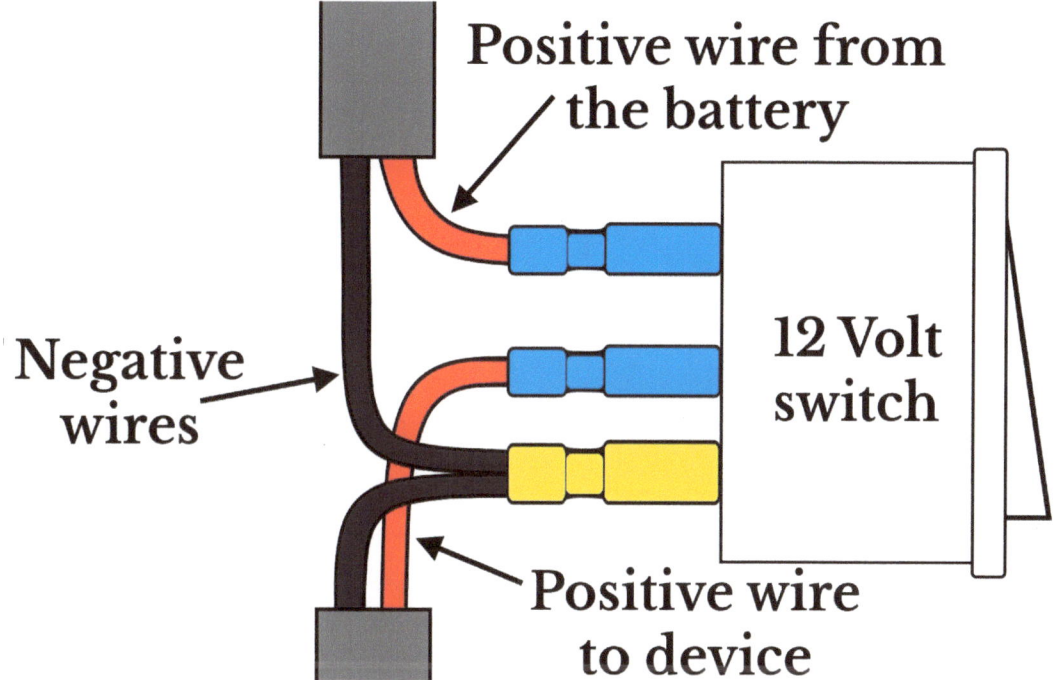

The three pin switch has three pins on it. Two pins are for the positive wires to allow power in and out of the switch like in the two pin switch design. The difference here is the fact that this switch won't work unless the negative lead from the battery and the device that you are powering are connected to the third pin. The negative pin is usually the brass coloured one whilst the two positive pins are usually silver in colour. A bigger wire connector (yellow) has been used for the earth wire to accommodate the width of both wires.

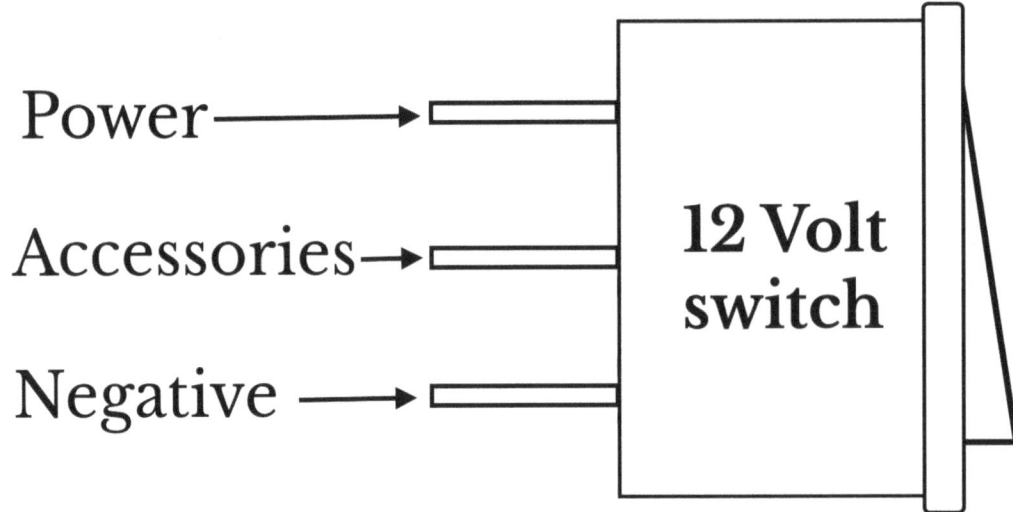

The power pin has the positive wire from the battery connected to it. The middle pin is usually called the "Accessories" pin. This is just another name for any devices that you intend to power with this switch.

The power comes into the first (Power) pin and then goes through the switch and out of the second (Accessories) pin to power your device (assuming that your switch is closed). And the third pin is purely for the two negative connections to be combined into one.

Switches do have an amp rating so ensure that you buy the switch that suits the amp rating on that circuit or simply buy the highest amp rated switches available (about 25 amp) and they should be good for most circuits.

These switches are sold under the name SPST which stands for Single Pole Single Throw. This means that if you push it one way it is off and if you push it the other way it is on. These switches are fairly cheap to buy so try and buy good quality ones with a fairly good amp rating so you are able to use them on various circuits as needed.

Chapter 9

Light bulbs

There is an array of different light bulbs that you can buy for your 12 volt system. EBay and Amazon are a good source for locating various types of these bulbs. The lower the wattage needed by the bulb to operate the less energy that you will drain from your batteries and the longer your lights will stay on.

Incandescent 12 volt light bulbs

The incandescent bulb is what most people are used to having in their homes. These bulbs whilst giving out plenty of light are quite wasteful in the amount of energy they consume. Generally speaking the incandescent bulbs that you are used to using in your house are bad news if you need to conserve energy on your 12 volt system. You can buy 12 volt 60 watt light bulbs that fit into a standard household fitting but these still use large amounts of power. Why would you use a 60 watt incandescent bulb when you can get the equivalent amount of light from an 8 watt (or lower) LED bulb?

There is a vast array of different incandescent light bulbs available for cars that can be readily adapted for use on your 12 volt system. These are not the most efficient type of bulb to use but they are readily available and fairly low priced. These can be adapted to light your shed etc, but again they use more power than the LED or fluorescent bulbs so I generally try to avoid them. You will find that if there is a 12 volt bulb designed to fit into a car then there is an LED version of the same bulb which will use a fraction of the power to run. So it will generally make more sense for you to look for the LED equivalent.

Fluorescent bulbs

Fluorescent bulbs are glass tubes filled with mercury gas. An electric current excites the gas making it emit light. Because these glass tubes contain mercury it is actually classed as a toxic device but so long as the glass tube stays intact it remains safe enough to use.

These are readily available in 12 volt versions online and reasonably priced. They have standard connectors that make connecting them up to existing light sockets very easy.

Although these bulbs tend to be more efficient than the incandescent versions they are still slightly power hungry compared to the LED bulbs. They are relatively cheap and easy to buy which is why most people like them but no matter what the makers claim for their lifespan I have found that they fail fairly easily and quite regularly, unlike the LED version.

LED bulbs

Light Emitting Diodes can run using much less power than incandescent or fluorescent bulbs, for this reason they are the ideal solution for the 12 volt power systems. These are very cheap to buy & run because of their efficiency.

There is a vast array of different types of 12 volt LED bulbs that you can buy for your lighting needs. They are even available to fit into standard household bayonet light fittings. This means that standard household fittings can be used in your setup.

G4 LED bulb: These are tiny little bulbs that can pack a lot of light inside them especially when you use a few of them together. They are very cheap to buy and usually come in multi packs. For the amount of power that they use they are a very good buy. They can be used in the place of drop lights very easily and will use a fraction of the power to run compared to conventional bulbs.

MR16 bulbs: These are the types of bulbs that are used in drop lights. They have two prongs for the positive & negative connections but generally speaking these bulbs are not too fussy which way around the pins go, the polarity doesn't generally matter a great deal with these bulbs. These bulbs are cheap and easy to buy from most online retailers. There is a halogen version of these bulbs that you need to avoid because they need much more power to run these.

LED light strip: These are rolls of tiny light emitting diodes all strung together onto one long strip. These are great for lighting under shelves and general background lighting. They don't usually give out enough light to work under unless you have lots of them but for the amount of power that they consume they are a very good lighting solution and give out a nice warm background lighting solution.

Corn light bulb: These are given this name because they look similar to a head of corn. These can fit straight into a normal house fitting which makes them a great solution for most people. They can be a little pricey compared to normal incandescent bulbs but they are worth it due to the reduced amount of power that they can consume. These bulbs are available in quite high lumen outputs which means that they will light up your room a treat.

LED light panels: These are actually design to fit to the interior light of your car and generally go behind a diffused piece of glass or plastic. They can however be used inside a shed etc because they give out a lot of light whilst using very little power. They can be a little harsh on the eyes however because they are so bright which is why they are housed behind defused glass. They are very cheap to buy and because they use very little power you can have quite a few of these on your ceiling.

How bright is a bulb?

The light that you get from a standard incandescent bulb was always measured by the amount of power that the bulb consumed in order to function properly. This is no longer the case because of newer more efficient bulbs that can now use much less power to produce the same amount if not more light. This is why a new scale to measure the light output of bulbs was created called Lumens.

As a guide to compare your favourite incandescent light bulbs to the modern equivalent here is a small conversion chart to allow you to buy bulbs that will give you the equivalent light output.

40 watt = 600 lumens

60 watt = 900 lumens

75 watt = 1125 lumens

100 watt = 1500 lumens

Generally speaking the LED bulbs are your best choice on your 12 volt solar power system. Search EBay and Amazon for ones that suit your needs best and start connecting them up to your system.

Chapter 10

Selecting the right wire for your circuits

We shall be sizing our wires using the AWG (American Wire Guide) measurements. This is a fairly standard measurement for wire which is used all around the world. This gives you the thickness of the wire and the recommended amp rating for each wire stated on the guide.

Before we start talking about different wire thicknesses, you need to understand one little quirk that the AWG measurements scale has. The thicker the wire is then the lower the number will actually be on this scale. So wire that has a thickness of "0" on the AWG scale is one of the thickest wires you can get and it is extremely thick. And the higher the number on the AWG chart the thinner the wire becomes. This can confuse people at first and was not created to make things easier for the general public.

There are two basic measurements that are usually needed when selecting wires to use on your 12 volt system. You need to ensure the wire will be thick enough to handle the power (the amps) that will flow through it and you will need to know how long the wire will have to be to reach the device. With these two measurements you can select your wires quite safely.

The length of the wire matters because the longer the wire is the more energy that will get lost on the way to the device that it is powering. This means a long wire will have to be even thicker to allow more energy to flow through it in order to replace the energy that is getting lost along the way. This energy loss is mainly caused by the resistance to the flow of electrons inside the wire itself.

The maximum amount of amps you should need to use on your 12 volt solar setup will be about 10 amps. This 10 amp demand will be for your 12 volt power sockets which are rated at 10 amps maximum. To this end the chart below is setup up to cater for your entire amp needs. The chart will give you any power rating between 1 and 10 amps which will power all your devices on your 12 volt system.

For this easy learn system we have included a chart that gives you your wire thickness based on these two measurements. Simply decide how many amps are required on the circuit you are building and how long the wire must stretch. The chart will tell you how thick your wire must be to remain safe. These figures have a 3% acceptable power loss factor.

Device Amp Usage	AWG wire size								
1 amp	26	23	22	20	19	18	18	17	17
2 amp	23	20	18	17	16	15	15	14	14
3 amp	22	18	17	15	14	14	13	12	12
4 amp	20	17	15	14	13	12	12	11	11
5 amp	19	16	14	13	12	11	11	10	10
6 amp	18	15	14	12	11	11	10	9	9
7 amp	18	15	13	12	11	10	9	9	8
8 amp	17	14	12	11	10	9	9	8	8
9 amp	17	14	12	11	10	9	8	8	7
10 amp	16	13	11	10	9	8	8	7	7
Wire length in meters	1	2	3	4	5	6	7	8	9

Example 1: A wire is needed that is capable of supporting a 3 amp bulb and the wire needs to be 4 metres long. Using the chart we first look at the amps needed and then slide across that column till we reach the 4 metre length at the bottom of the chart. This tells us that a 15 AWG wire will do the job nicely.

When measuring a wire to power this light bulb in the centre of the ceiling you would have to run it from the battery all the way up the wall and then to the centre of the ceiling itself. Even in a small shed this is usually about 4 metres in length. This wire does actually loop back on its self via the negative wire but for our calculation we only need to measure it one way.

Example 2: A 10 amp power socket with a 4 metre cable would need a much thicker cable to be safe. Using the previous chart you can see that it would actually need the much thicker 10 AWG wire to be safe. You can use different thickness cables on each circuit or you can simply work out the thickest wire that will be needed on you system and buy one roll of this wire to use on all your circuits. This may be a bit more expensive in the long run but it is so much easier to do and your system will actually look better because everything looks the same.

Small note: Using the chart we worked out what a suitable wire would be for your circuit but if you wish to be extra safe you can always go for the next thickest wire size that is available to you. If you need the AWG 10 gauge wire for a circuit then you could always choose to buy the next thicker wire (AWG 9) so that your circuit is even safer. This isn't necessary but if you can afford it then it will always be the safer option.

Chapter 11

Inverters

Having a 12 volt system is great but what if you need to power a mains powered device such as a vacuum cleaner or a power tool. Your 12 volt system will not be able to make these devices work unless you use a device called an inverter.

An inverter in case you have never heard of one takes 12 volt power from your battery and scales it up to mains voltage electricity so that you can power your televisions, games consoles and power tools etc.

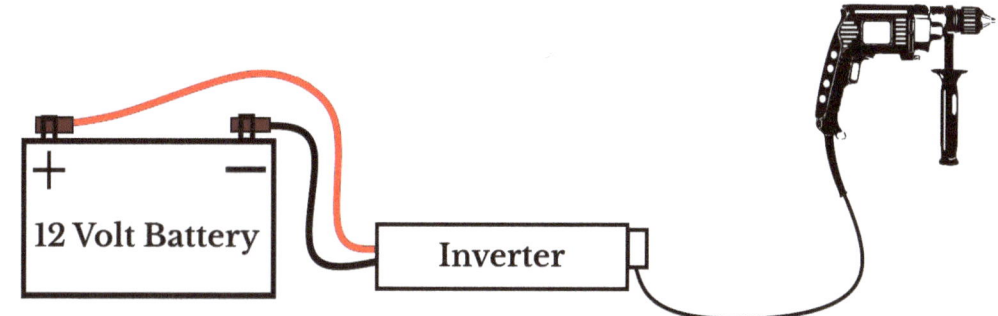

12 Volt battery + Inverter = Power for TV's, Consoles, Power tools etc

They are ridiculously easy to use because all you have to do is attached the clamps that come supplied with it to the terminals on the battery, plug in your AC mains device, power up the inverter and then you can switch the device on. This means that you don't have to use just 12 volt devices when you are off grid because you are able to use most mains powered devices (within reason) so long as the inverter is powerful enough to support the devices needs.

Inverters come with the amount of watts they can power written on them. So if your mains powered device needs 1000 watts of power to function properly then you will need a 1000 watt inverter as a bare minimum. You will find however that buying a 1500 watt inverter would be a wiser purchase as it's always better to have it larger than needed this way you won't overtax an inverter that is too small for the task.

There are two types of inverters on the market for you to buy. The **Modified sine wave** and the **Pure sine wave inverter**.

The cheapest type of inverter is the "**Modified sine wave inverter**". This type of inverter usually works fine on most general electrical stuff such as power tools and televisions. Certain devices can however be a bit temperamental when being powered by these and either refuse to work properly or they may give off strange noises whilst in operation. Devices such as rechargeable batteries and laptops can also dislike these. This is because the electrical wave form generated by these isn't quite the same as the one generated by the mains power supply itself. It literally chops the top and bottom off the wave form that it generates making it more square in shape rather than the more rounded shape that is produced by a mains supply. This can make some devices uncomfortable in operation.

The more expensive "**Pure Sine Wave Inverter**" produces a wave form that is identical with that generated by the mains electricity itself, so devices love them. The only problem is that these inverters tend to cost much more to buy because of the extra time and higher quality parts that are needed for their construction.

Generally speaking the cheaper modified inverter will suit most of your needs but if you intend powering delicate electronic equipment such as laptops then the pure wave inverter may be your best bet.

The inverter should be positioned as close to the 12 volt battery as possible so that as little energy as possible is lost powering it. Always use an extension lead to power your mains powered devices as this means that you can have mains power anywhere you want it to be. Never be tempted to extend the leads on the inverter so that you can have it in a more convenient position. Twelve volt power does not travel over (even small) distances very well, the longer the lead on a 12 volt device the thicker the cable has to be, and the more power you lose on the way. AC power doesn't have the same problems as DC power when it comes to travelling longer distances. This is why you should use an AC extension lead to bring this power to your AC mains devices wherever they are situated.

Small warning: Always clamp the inverter to the terminals of the battery with the thickest set of leads that come with the inverter in order to power it. Avoid powering it via the 12 volt power socket plug that comes with it. It is always better to have a direct connection with the battery rather than having all the power going through a 10 amp power socket in a vehicle or on your solar setup. The inverter can draw a lot of power at any one time which can overload the thinner power socket lead that comes with it, unless you are only powering a very small device.

Chapter 12

Installing it & connecting it all up

Using the ultra safe & easy basic circuit on your 12 volt system means that you will find the fitting of the wires & the devices a very simple process. There are however two different setup variations that you can use. The first diagram shows the power coming from the load section of the charge controller and the second diagram shows it coming directly from the battery itself. Take a look and see which version you prefer.

The Load section style setup

The solar panel is connected to the charge controller so that it can manage the incoming power. The charge controller is connected to the battery so that it can charge it up.

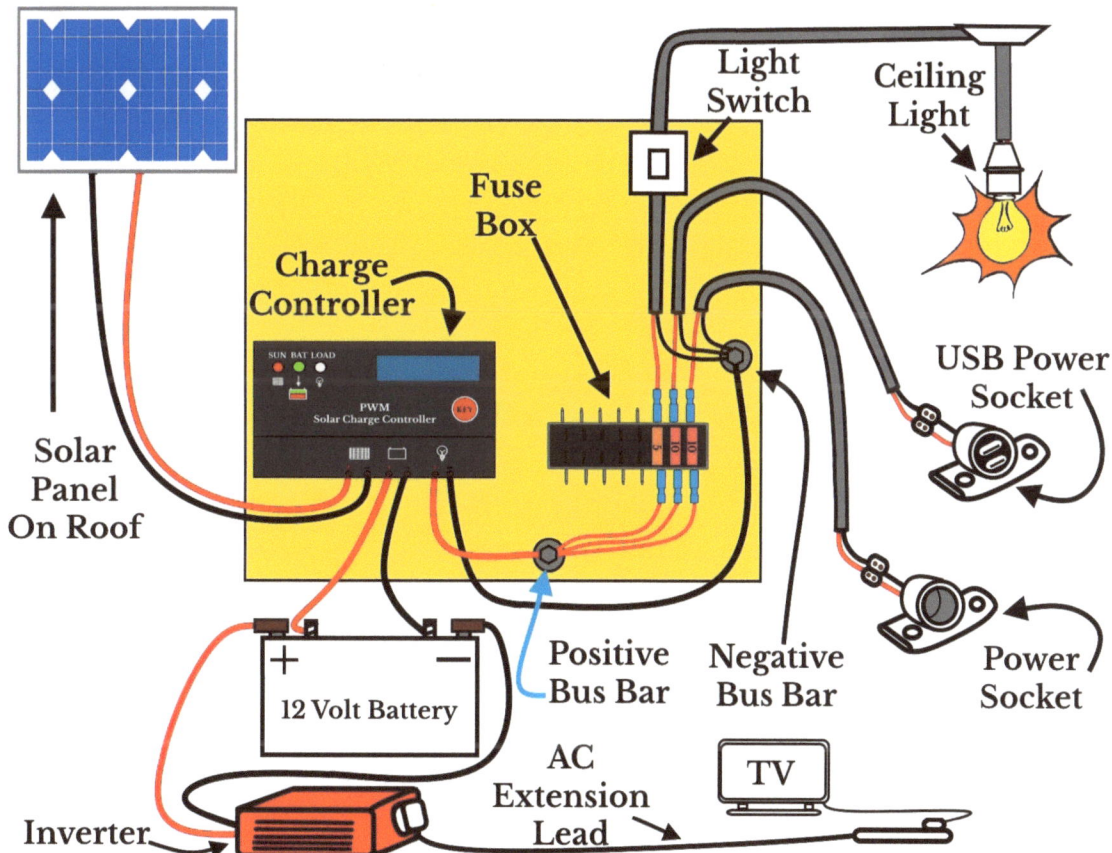

The power from the load connection on the solar charge controller is attached to a bus bar (nuts & bolts in this case) that in turn distributes all the power to each individual circuit on your fuse box. Because the power is all coming from the load section of the charge controller you will be limited on how many

37

circuits that you can have running on it due to the amp rating of the charge controller itself. The charge controller in this case would be rated at 30 amps so this circuit has been setup so it cannot exceed 25 amps in total.

The fuse box creates the safety that your circuits need and then allows all the power to travel up each individual circuit wire till it reaches the device. The power travels through the device and then returns to the negative terminal on the battery (via a negative bus bar).

The battery supply style setup

If you were to power your circuits directly from your battery terminals as shown in the next diagram then you would not have to worry about exceeding the limits of the charge controller because all the power comes directly from the battery itself cutting out the charge controller's part in powering the circuits. This would mean that you could add extra circuits when needed. The battery fed system should look similar to the diagram below.

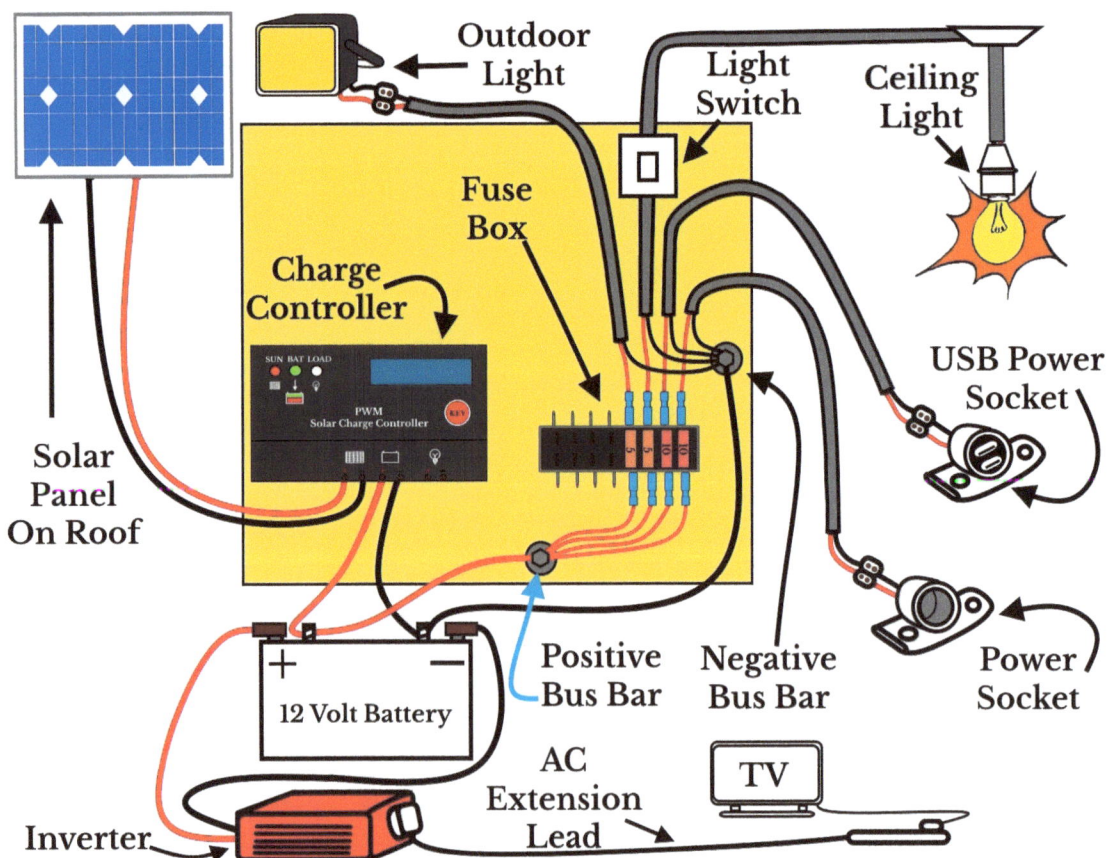

You can now see both ways of wiring up your system it is up to you to choose if you wire it straight from your battery or via the load option on the charge controller.

The charge controller load option has a power management feature depending on what mode the charge controller is set on. You can have it set so that it switches off all the power to your circuits if the power in your batteries drops below a certain point, thus protecting your batteries. If you use this option however you are limited to the amp rating of the charge controller for the amount of circuits that you can power in total. I personally prefer to power my circuits straight from the battery but this decision is yours.

Small note: This electric control panel has been shown with only one switch displayed and this switch is actually the light switch which can be placed anywhere in the room. What hasn't been shown is the fact that it is always better to have a switch on each circuit straight after the fuse box. Having a switch on the control panel for each circuit gives you more control over your system and enables you to switch various circuits on or off as needed from one central location which is ideal in an emergency situation or whilst performing maintenance on a circuit. Due to lack of space in this diagram these have been omitted but it is recommended that you do fit a switch on each circuit when building yours.

Small note: No voltmeter has been shown in these diagrams but you are advised to fit one so that you can monitor the power levels in the battery.

Chapter 13

Series & parallel circuits

As promised here is the chapter on series and parallel circuits. This is just a general overview of how they are constructed and how they work. This chapter has been added so you understand these circuits and how they work but for someone with absolutely no experience using electrics I would recommend that you stick to using just the basic circuit design for now until you feel more confident in what you are doing.

The series circuit

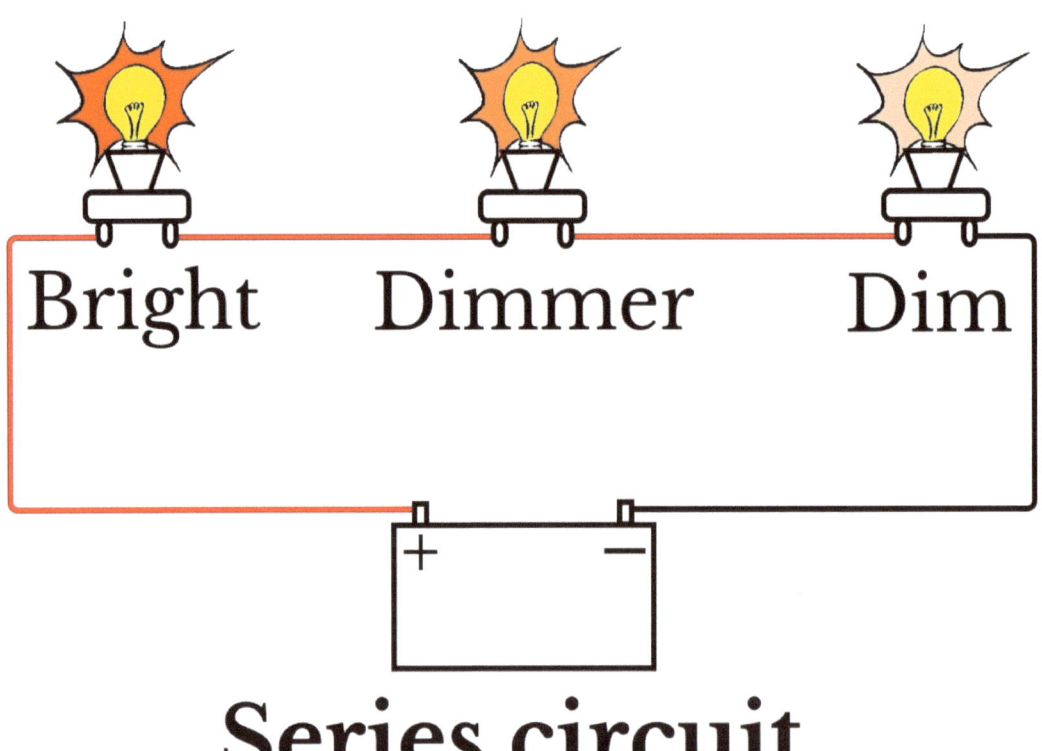

The power comes from the positive terminal and moves through each light bulb in series. It has to go through one bulb before it is able to go through the next one and then the next one before it can return to the negative terminal and the battery.

The first bulb is nice and bright because it gets all the power it needs to light up properly. The second bulb doesn't shine quite as bright because some of the power has already been used up by the first light bulb. The third light bulb however is really struggling to even light up at all because the first and second bulbs have taken most of the power and as a result it shines quite dimly. This is the first major problem with a series circuit.

The next problem with the series circuit is the fact that if one bulb was to burn out it would create a break in the circuit similar to the way that a light switch or a fuse does and stops the power from flowing in the circuit itself. The power would reach the broken bulb and would then find itself unable to move any further until the bulb itself was replaced. This means that all the bulbs would stop working even though only one bulb was faulty.

It isn't difficult to work out how many amps a series circuit uses because all the amp requirements of the devices on a series circuit use are simply added together to get the total amperage that will be needed for the devices on that single piece of wire. If you have two 3 amp bulbs on a series circuit then the total amount of amps the wire will have to carry will be 6 amps, but you always need to make your wire thicker than this for safety reasons so the wire needs to be able to cope with 8+ amps as a general rule of thumb.

This setup can work but it isn't ideal because of the way the power struggles to get through the circuit itself. Because of this loss of power the voltage on the circuit itself will drop by the time it gets through to the other side of the circuit, meaning that although it started off as a 12 volt power source it won't be 12 volts by the time it gets to the negative terminal.

This voltage drop can cause problems and even cause fires in extreme cases which you really don't need, so it is better for you if you do not use the series circuit if possible. The basic circuit you were shown first is the safest option for you at the moment.

Technically speaking the basic circuit is a series circuit because there is only one path for the power to get through, but because you are only powering one device there should never be a problem with the amount of power supplied to the device on that circuit.

Parallel circuits

The parallel circuit is fairly easy to understand because all you are doing is giving the power multiple path ways to go through. Instead of only being able to travel through one wire like we have just seen in the series circuit, it now has multiple wires that it can travel across. The power spreads itself out and travels down each of the wires so that it can pass through the bulbs and get back to the battery.

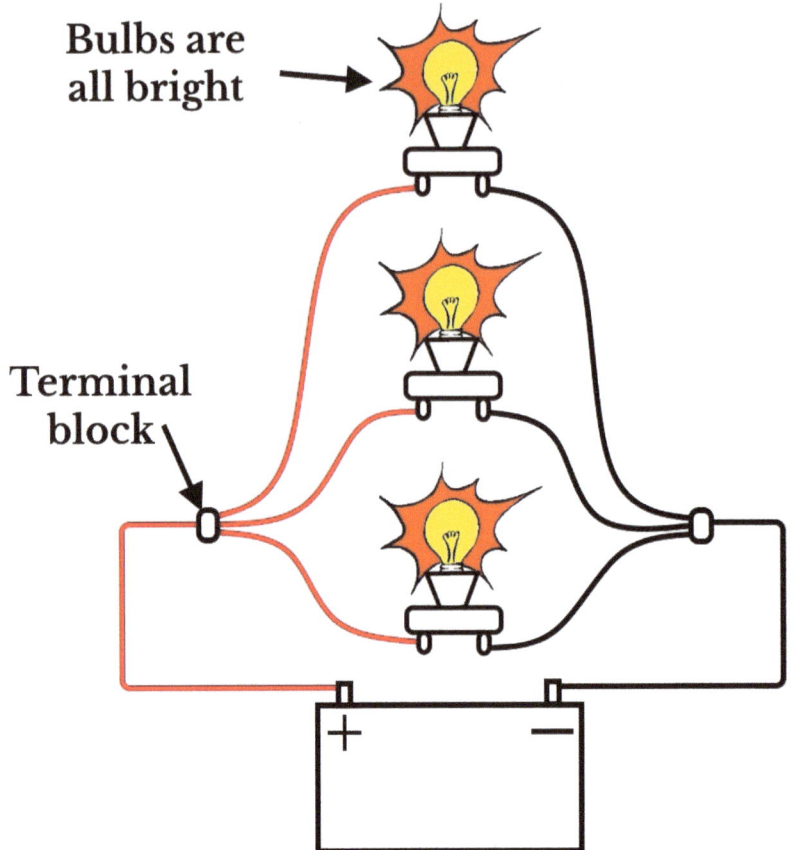

Parallel Circuit

In the parallel circuit the power doesn't have to travel through each bulb or device one after the other. The power is able to travel through any device on the circuit that it wishes in order to get back to the negative terminal. This means that if one bulb stops working the power can still get back to the battery by travelling through the other bulbs. This also means that you don't get dim bulbs because the power reaches all of the bulbs and each bulb can take the amount of power that it needs to shine at the correct brightness.

Unlike the series circuit you don't get a voltage drop in a parallel circuit because power can travel down many various pathways to get back to the battery. The 12 volt power source will stay constant and all the devices should hopefully get an adequate amount of power. You will find that you will actually get a greater amount of current flowing through the circuit because there are so many different routes for the power to travel through.

This makes the parallel circuit a much better choice for you if you intend powering multiple devices from a single length of wire. However to use it on your 12 volt setup you must fully understand how to calculate its power & fuse rating requirements.

The problem with parallel circuits is the fact that you cannot simply add together the different amp ratings to get a safe wire & fuse rating for the circuit like you do with a series circuit.

Working out the resistance on a parallel circuit can involve some reasonably intense maths which most people will struggle with. First you will have to convert the amp rating of each device into its resistance value by dividing the voltage by the amp rating.

(Voltage) V÷I (Amps) = Ω (Ohms) –Ohms being the term that describes the resistance of a device.

Then you have to take the resistance of each device and convert it to a fraction, add these together and convert this to the reciprocal to get the total resistance on the circuit.

I personally have never liked maths so I don't do this, I cheat. I place one large fuse in the fuse box to cover the estimated total amperage that this circuit may need to support and then I place a smaller individual fuse just before each device that is matched to the needs of the device itself. Just like your television set that has a built in fuse in its plug to protect it, my circuit has a fuse rated purely for the device itself and then the larger one in the fuse box is designed to protect the whole circuit. Doing this means that I don't have to do any complicated maths equations.

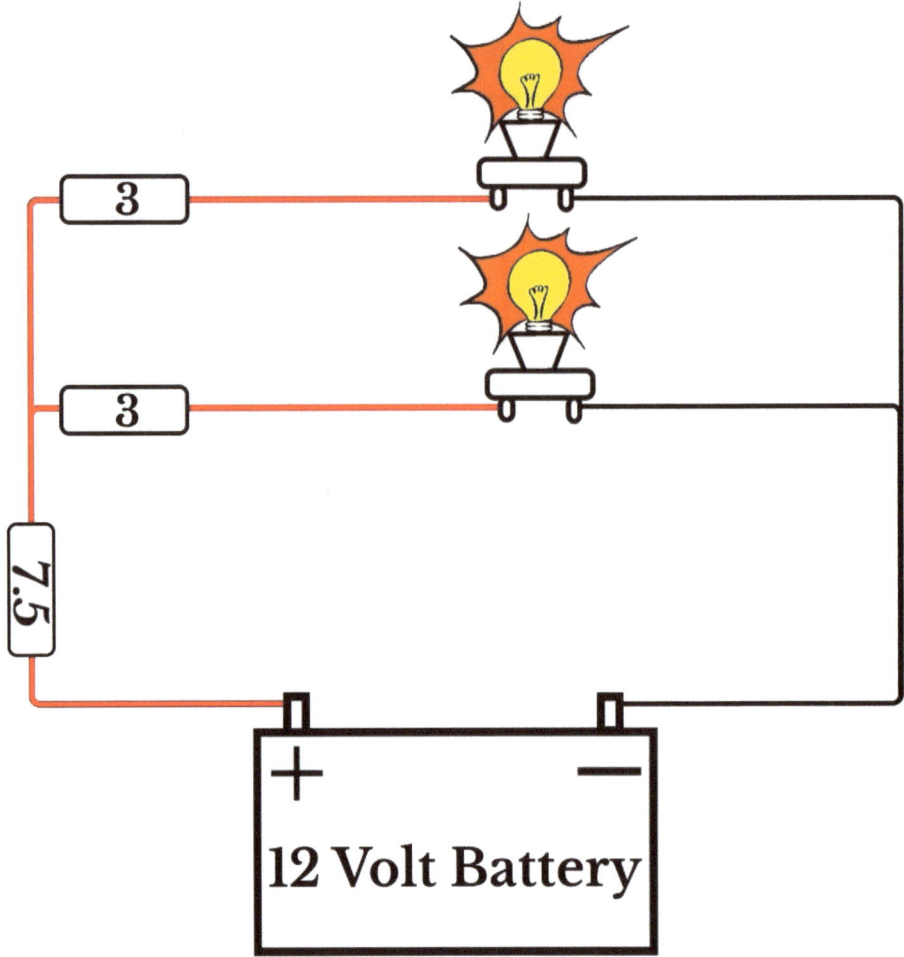

I would have preferred to have used a 6 amp fuse in the above diagram but unfortunately they don't make 6 amp fuses in the blade type fuses so the 7.5 amp fuse was the next closest one available for me.

As a general rule I only ever keep each individual circuit purely for one type of device i.e. circuit one is for lights, circuit two is for power sockets and circuit three is for external lights. I don't like to mix and match different power rated items on any one circuit. This helps to ensure that one device isn't under or overpowered by another devices power demands. The last thing that you want is the power to try and push its way through a device with a lower resistance rating than another device because it is easier for it to get through that way.

I also generally limit my parallel circuits to just two devices per circuit. This is because the power splits fairly evenly on a circuit with two devices, whereas circuits with more than two devices start to get a little fuzzy in the way that it shares the power out. You can put more than two devices on a parallel circuit but I personally prefer sticking with just two devices per circuit.

Parallel circuit terminal blocks

This extra section has been included in order to show you how to create any parallel circuit you wish using a connector block, which will make constructing them so much easier for you if you decide to use them.

You will find that terminal blocks are ideal for wiring up your parallel circuits whenever you decide to use them. The beauty of these connectors is that you have the ability to modify them so that you can attach the positive & negative wires into one hole and then join multiple devices to your main wire quickly and easily (sounds confusing until you look at the diagram). You can easily power multiple bulbs from just this one wire. Simply by adding small loops of wire into the neighbouring holes on the terminal block. The power is able to travel through any connection that has a loop linking it.

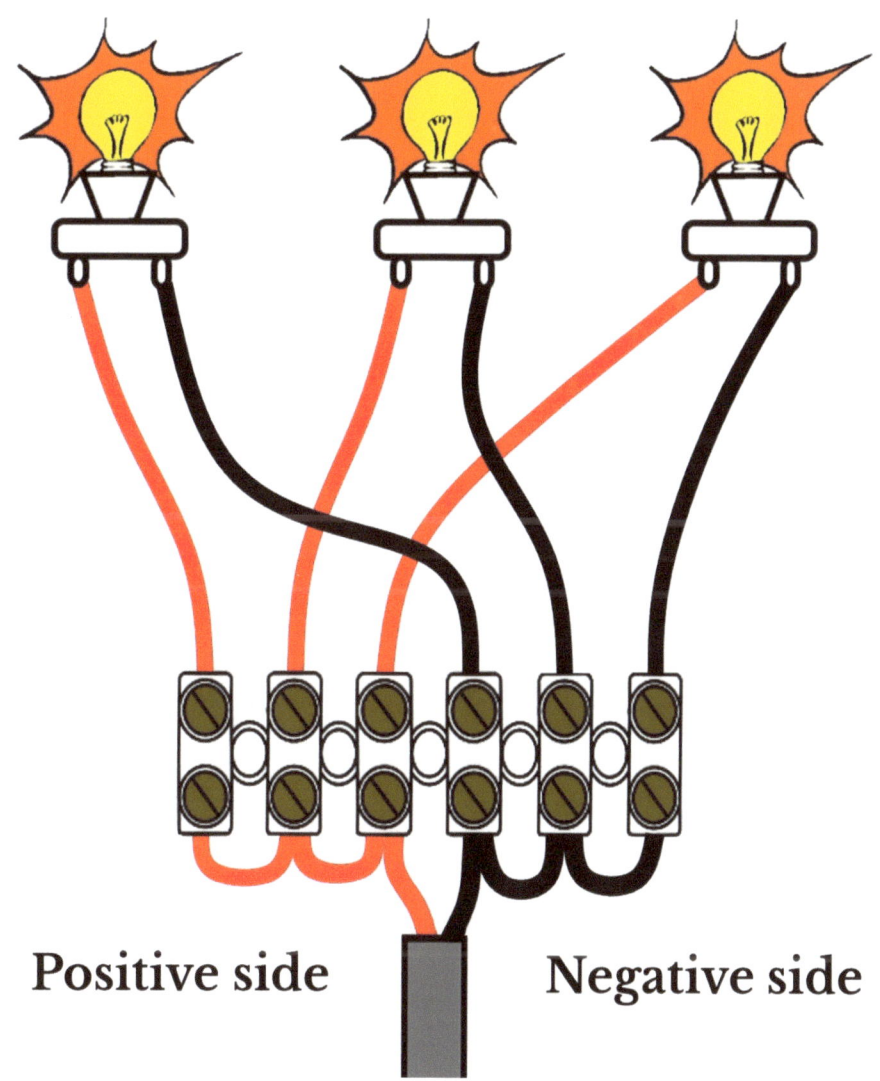

Positive side   Negative side

The bulbs & the wires attached to them in this picture have been shrunk down quite considerably so they would fit into view properly. In reality these bulbs & wires would be stretched out across your ceiling. The power flows into the circuit via the red positive wires, goes through the light bulbs and returns back to the battery via the black wires.

The diagram does show you that by simply looping small lengths of wire into the adjacent holes next to the positive & negative wire (the copper wires must be in contact with each other) the entire terminal block turns into a parallel connection point. Power can flow through the first bulb and get to the other side but it can also travel along the small loops of wire and get to every other bulb on the circuit without having to travel through just one bulb or wire. If any bulb in this circuit stops working the power can simply travel down any of the other loops of wire in order to get back to the negative wire and the battery.

Playing it safe

If you are in any doubt about how to calculate the amp ratings on different types of circuits such as the series & parallel types then don't bother using them, simply use the basic circuit design shown at the beginning of this chapter. The basic circuits are much easier to build & understand, and they work brilliantly. There is no shame in sticking to the basic design of circuit for your power needs. It is better to create four or five super safe basic circuits to power your different devices than to have one or two series or parallel circuits that may not have been safely calculated for their amp or fuse rating. These basic circuits are quick to build and easily rated for fuses and best of all they are super safe to use.

Chapter 14

Goodbye

Good luck on installing your 12 volt off grid solar system. Remember to take your time and measure out the lengths of cable to suit the amount of amps that you will be using on that wire.

As you install your own devices you will see how easy it is and your confidence levels will soar. You may even be tempted to increase your knowledge of circuits etc and progress onto bigger and more powerful systems. This is all down to you now. You have been shown all the basic information you require to get a 12 volt solar power system up and running, the rest is all down to you.

Good luck & goodbye!

www.ingramcontent.com/pod-product-compliance
Lightning Source LLC
Chambersburg PA
CBHW051219220526
45473CB00003B/1101